Advancing Resilient Performa

Christopher P. Nemeth • Erik Hollnagel
Editors

Advancing Resilient Performance

Editors
Christopher P. Nemeth
Applied Research Associates, Inc
Albuquerque, NM, USA

Erik Hollnagel
Macquarie University
Sydney, Australia

ISBN 978-3-030-74691-9 ISBN 978-3-030-74689-6 (eBook)
https://doi.org/10.1007/978-3-030-74689-6

This Springer imprint is published by the registered company Springer Nature Switzerland AG
The registered company address is: Gewerbestrasse 11, 6330 Cham, Switzerland

To the frontline workers who daily anticipate what may be needed and create the resilient performance this text advocates.

Contents

Contributors

Matthew Alders King's College London, London, UK

Janet E. Anderson King's College London, London, UK

Asher Balkin The Ohio State University, Columbus, OH, USA

Pierre Bérastégui University of Liège, Liège, Belgium

Meredith Carroll Florida Institute of Technology, Melbourne, FL, USA

Alexander Cedergren Lund University, Lund, Sweden

Pedro Ferreira CENTEC – University of Lisbon, Lisbon, Portugal

Erik Hollnagel Macquarie University, Sydney, Australia

Henrik Hassel Lund University, Lund, Sweden

Sudeep Hegde Clemson University, Clemson, South Carolina, United States

Cullen D. Jackson Beth Israel Deaconess Medical Center, Harvard Medical School, Boston, MA, USA

Akinori Komatsubara Waseda University, Tokyo, Japan

Beth Lay Lewis Tree Service, West Henrietta, NY, USA

Shem Malmquist Florida Institute of Technology, Melbourne, FL, USA

Christopher P. Nemeth Applied Research Associates, Inc, Albuquerque, NM, USA

Anne-Sophie Nyssen University of Liège, Liège, Belgium

Gesa Praetorius Linnaeus University, University of South-Eastern Norway, Notodden, Norway

Anne Marie Rafferty King's College London, London, UK

Eric Rigaud MINES Paris Tech, PSL-Research University, CRC, Paris, France

Mark Sujan Human Factors Everywhere, Ltd., Woking, London, UK

About the Editors

Erik Hollnagel is visiting professorial fellow, Macquarie University (Australia); and visiting fellow, Institute for Advanced Study, Technische Universität München (Germany). He is also professor emeritus from Linköping University (Sweden), Ecole des Mines de Paris (France), and the University of Southern Denmark. Erik has throughout his career worked at universities, research centers, and with industries in many countries and with problems from a variety of domains and industries. He has published widely and is the author/editor of 26 books, including 6 books on resilient healthcare, as well as a large number of papers and book chapters.

Christopher P. Nemeth conducts human performance research and development in high-hazard sectors as a Principal Scientist with Applied Research Associates, a 1500-member US science and engineering consulting firm. His 26-year academic career has included 7 years in the Department of Anesthesia and Critical Care at the University of Chicago Medical Center, and adjunct positions with Northwestern University's McCormick College of Engineering and Applied Sciences, and Illinois Institute of Technology. He has served as a committee member of the National Academy of Sciences, is author/editor of five books, and is widely published in technical journals.

List of Abbreviations

AMU	Acute Medical Unit
FAA	Federal Aviation Administration
FPS	Fatigue Proofing Strategies
FRAM	Functional Resonance Analysis Method
FRS	Fatigue Reduction Strategies
HOS	Hours of Service
HSIB	Healthcare Safety Investigation Branch
LFI	Learning from Incidents
NASA	National Aeronautics and Space Administration
NHS	National Health Service
NRLS	National Reporting and Learning System
NTSB	National Transportation Safety Board
PRIMO	Proactive Risk Monitoring
QoWL	Quality of Work Life
RAG	Resilience Assessment Grid
RE	Resilience Engineering
SOP	Standard Operation Procedures
WaI	Work as Imagined
WaD	Work as Done

From Resilience Engineering to Resilient Performance

Erik Hollnagel and Christopher P. Nemeth

Contents

Looking back at how resilience engineering made its entry onto the safety arena in 2004 (Hollnagel et al., 2006), it is tempting, and probably not completely misleading, to see Perrow's proposal of normal accidents as one of the conceptual precursors. Perrow's concern was the accidents that could occur in complex industrial systems, and his argument was that "(m)ost high-risk systems have some special characteristics, beyond their toxic or explosive or genetic dangers, that make accidents in them inevitable, even 'normal'." (Perrow, 1984, p. 4). The special characteristics that made "normal accidents" inevitable were identified as the interactive complexity and tight coupling that resulted in nonlinear interactions, defined as "unfamiliar sequences, or unplanned and unexpected sequences, … either not visible or not immediately comprehensible" (Ibid, p. 78). The introduction of the concept of normal accidents became a catalyst for a gradually growing dissatisfaction with the traditional concept of safety, even though it until then had served industries and societies well.

The subtitle of Perrow's book – *Living with High-Risk Technologies* – made clear that his concern was the high-risk systems that we both had to live with and had made ourselves dependent on. This was in good agreement with the tacit acceptance that safety was about how failures of technologies and systems could lead to

E. Hollnagel
Macquarie University, Sydney, Australia

C. P. Nemeth (✉)
Applied Research Associates, Inc., Albuquerque, NM, USA
e-mail: cnemeth@ara.com

C. P. Nemeth, E. Hollnagel (eds.), *Advancing Resilient Performance*, https://doi.org/10.1007/978-3-030-74689-6_1

unwanted and unacceptable outcomes; to accidents and incidents. When the notion of resilience began to appear in safety discussions, roughly around the turn of the century (Woods, 2000), the traditional safety interpretation still dominated, although there was a slowly growing realization that something was amiss. The conventional understanding of safety, for example, implied a hypothesis of different causes in the sense that the causes of adverse events had to be different from the causes of events that went well (Hollnagel, 2012a, 2012b). Otherwise, the elimination of such causes and the neutralization of the "failure mechanisms" would also reduce the likelihood that things could go well, hence be counterproductive. This dilemma was made clear already in the first book on resilience engineering (Hollnagel et al., 2006), which argued that:

> ... failures are the flip side of successes, meaning that there is no need to evoke special failure mechanisms to explain the former. Instead, they both have their origin in performance variability on the individual and systemic levels, the difference being how well the system was controlled.
>
> It follows that successes, rather than being the result of careful planning, also owe their occurrence to a combination of a number of conditions. While we like to think of successes as the result of skills and competence rather than of luck, this view is just as partial as the view of failures as due to incompetence or error. Even successes are not always planned to happen exactly as they do, although they of course usually are desired – just as the untoward events are dreaded. (p. xi)

A consequence of this change in perspective was that safety concerns no longer should be limited to high-risk systems but should also include systems of the more mundane kind. The production, service, and health care sectors are examples, but Perrow's worries unfortunately apply to these as well. His concern was that growing interactive complexity and tight couplings could lead to nonlinear interactions. These could lead to "unfamiliar sequences, or unplanned and unexpected sequences that were either not visible or not immediately comprehensible." The problem was – and remains – the rapidly increasing complexity of technologies and societies, made worse by our general willingness to make ourselves dependent on systems that are partly incomprehensible and, therefore, impossible to fully control. Problems are usually answered by patches; temporary solutions that only provide a short-term relief, and they are an unfortunate consequence of our inability fully to understand how these systems function.

A proper solution to this problem must clearly include how a reliable functioning of these systems can be ensured, hence go beyond the conventional efforts to prevent failures and reduce risks. There was then, and is now, a need to evolve from safety in the traditional interpretation, to move beyond the link between accidents and safety, and to find a concept that corresponds better to the new reality. Twenty years ago, resilience was suggested as a possible solution, and the suggestion was eagerly welcomed by the safety community.

In the beginning, resilience was defined as "the intrinsic ability of an organisation (system) to maintain or regain a dynamically stable state, which allows it to continue operations after a major mishap and/or in the presence of a continuous stress" (Hollnagel, 2006). This definition reflected the historical context by contrasting two states: one of stable functioning and one where the system has broken down. Following the legacy of industrial safety thinking, the definition was also limited to consider situations of threat, risk, or stress. Indeed, for many years resilience in yet another juxtaposition was defined as the antidote to brittleness.

Five years and several books later, the definition of resilience had changed to "the intrinsic ability of a system to adjust its functioning prior to, during, or following changes and disturbances, so that it can sustain required operations under both expected and unexpected conditions" (Hollnagel, 2011a, 2011b). In this definition, the emphasis on risks and threats had been reduced, and the reference instead became how systems performed in "expected and unexpected conditions," including how such conditions could be anticipated. The focus had also changed from safety criticality and responses to unplanned and unexpected sequences. The focus is now on the ability to perform or function as required, not only in the face of adversity but more importantly during normal conditions as well. Today, ten years later, a working definition of resilience might be the ability to succeed under varying conditions, so that the number of intended and acceptable outcomes (in other words, everyday activities) is as high as possible.

The change in the definitions since 2006 has served to broaden the scope of resilient performance. It is no longer just the ability to recover from threats and stresses, but rather the ability to perform as needed under a variety of conditions – which means being able to respond appropriately to both disturbances and opportunities. The inclusion of opportunities signals a change from protective safety to productive safety. Safety is no longer a cost but is instead an investment. Ultimately, resilience should be dissociated from safety, thereby leaving the increasingly sterile discussions and stereotypes of the past behind. Resilience is about how systems perform, not just about how they remain safe. Indeed, systems that are unable to make use of opportunities are not in a much better position than systems that cannot respond to threats and disturbances; at least not in the long term.

Another way in which the definition has changed is that it now is about the characteristics of resilient performance rather than about resilience per se. Resilience is not a mystical or mythical system property or quality as such, or something that can be measured or managed on its own, and therefore not something that can be engineered either. This was actually made clear from the very start, although it did not attract much attention. In the *Epilogue* of the first book, it was argued that safety, and therefore *a fortiori* resilience, was something that a system *did* rather than something that it *had* (Hollnagel & Woods, 2006, p. 347). Or as David Woods later put it more forcefully: "Resilience is a verb" (Woods, 2018). (Grammatically it would be better to say that "resilient is an adjective.") Resilience engineering as a field of research and practice is consequently about the characteristics of resilient performance, how we can recognise it, how we can assess (or measure) it, how we can improve and advance it. The *Epilogue* again tried to make that clear by pointing out that "(w)e can only measure the potential for resilience but not resilience itself" (op. cit).

The book you have now started to read is, therefore, not about safety in the usual meaning of the term, and possibly not even about safety at all. Neither is it about resilience and the engineering thereof. It is rather about understanding work and understanding how socio-technical systems perform, so that we can make sure that they will also work the next time around. This is actually very important for safety in the conventional sense, since understanding how things work is also the basis for understanding how outcomes can vary. Failures *are* the flip side of successes, and can, therefore, not be understood separately from them.

A list of the issues addressed by the various chapters clearly shows that advancing resilient performance is about work rather than about safety and, therefore, about coping with complexity rather than recovering from failures. The issues cover how to use rules wisely, fatigue management, work-as-done in an acute medical unit, limitations in learning from incidents, ways to share information and experience, how to enhance the potentials for resilient performance, autonomous maritime operations, how to increase people's ability to handle variability, addressing problems of structural secrecy, and how to clear trees from power lines. Much of the content is clearly relevant to ensure that performance takes place without unplanned and unexpected developments and outcomes. This is also a consequence of safe performance in the traditional sense, but in this case incidental rather than deliberate. Resilience *is* a verb. Resilience is about how things *are being done* and about how we can find ways to manage and advance resilient performance.

1 Trends in Applications

Christopher Nemeth and Erik Hollnagel

The central questions in this book are what resilient performance means and what can be done to advance it. Work in the earlier years of resilience engineering centered around observation and the identification of work as it is actually done, in contrast to work as it is imagined. Many authors initially embraced the venerable tradition of field research known from countless workplace studies (Luff et al., 2000). However, the interest of resilience engineering in adaptive capacity, interdependencies, and development of solutions grounded in understanding work-as-done moves beyond workplace studies. This has resulted in new knowledge and useful methods that have demonstrated the value of this approach, in addition to, or as a replacement for, current practices (Nemeth & Herrera, 2015).

Two chapters address themes that can be applied across application areas. First, Komatsubara examines the role that a manager plays to guide workers on when to follow rules and when to deviate from them. The chapter frames resilient performance in simple terms, translating more abstract descriptions into practical terms that a manager can use to guide workers by showing how to accomplish production safely. Rather than separating Safety-I from Safety-II (Hollnagel, 2014), he contends they belong together by establishing requirements and then complementing them with guidelines to manage them flexibly. This leads to a description of three kinds of "manuals," or procedures going from technical regulations that compel strict adherence, via rules that can allow deviation in emergency situations, to guides that invite change within acceptable norms. Workers vary in their ability to engage actual circumstances successfully. Because managers know their workers as well as which practices may be suited to conditions as they occur, they can play an essential role in ensuring safe production.

Bérastégui and Nyssen assess how fatigue affects performance, noting that imposition of strict external norms and practices without regard for actual operations can actually degrade performance. They present a case for a fatigue risk management system (FRMS) to find, develop, and implement risk-related procedures that can be tailored to an organization. The measurement of safety performance indicators makes it possible to determine the effectiveness of procedures that are identified by both fatigue reduction and fatigue proofing strategies. They conclude with a case study showing the use of FRMS in a Belgian hospital's emergency department and point to potential for it in other sectors, such as emerging "gig economy" ride hailing businesses. The recognition of worker insights as they engage workplace uncertainty can yield more robust means to identify and manage risk and a more grounded, flexible way to manage performance.

Six chapters study particular applications and distil principles that have a potential use beyond the particular area that were studied. Three study healthcare, which has been an area of interest from the outset, pioneered by practitioner/researchers including Cook (Cook & Woods, 1996; Cook et al., 1998), and Wears et al., 2006). The successful application of resilience engineering principles to the thorny issue of patient safety has already become an active field of study in itself (e.g., Hollnagel et al., 2019).

Alders, Rafferty, and Anderson translated Hollnagel's (2011) Resilience Assessment Grid (RAG) potentials (anticipate, monitor, respond, and learn) to improve the performance of an acute medical unit of a London hospital. Focus groups with nurses yielded a grounded description of work-as-done including their adaptations and adjustments. A 37-item survey based on the focus group findings polled staff members on how well each item was performed (e.g., "taking action to reduce workload for the next shift" as an anticipating potential). Semi-structured interviews on the results with seven of the focus group participants yielded interventions to improve unit performance. The approach ensured the analysis reflected actual work in the unit, produced a series of improvement recommendations aligned with each of the four RAG potentials, and revealed interdependence among units in the hospital.

Sujan used resilience engineering over 18 months to improve current healthcare operations at a radiology department of one National Health Service hospital in the UK, and in a surgical emergency admissions unit of another NHS facility. His approach is based on the argument that learning from incidents misses how healthcare professionals actually deliver care and fails to translate findings into learning. A resilience engineering approach would instead understand every day the trade-offs and adaptations that clinicians make and learn how they might be supported. Sujan describes a Proactive Risk Monitoring (PRIMO) approach to organizational learning, including staff narratives that describe problems, contribution of free-text narratives, short-term as well as long-term improvements, and staff ownership of interventions to improve performance. Generating interventions from within an organization increases the likelihood of successful implementation. Learning from work as it is performed encourages awareness and discussion within and among departments.

Hegde and Jackson describe their efforts to implement the Resilience Engineering Tool to Improve Patient Safety (RETIPS), a self-reporting system for hospital caregivers. The online format sought to make it easy to share accounts of adaptation in the workplace, reflecting actual work as it is performed. The interface invited selection of a case, entry of a narrative, a description of what went well, challenges and concerns, resources the contributor used, and area of clinical practice. Examples in the chapter demonstrate results collected using the pilot RETIPS system. The authors' candid reflection includes outcomes they did not anticipate, and provides insights into the complexities of change management. These insights included the need for buy-in from key stakeholders, connecting RETIPS to processes that already exist, providing an incentive to use it, maintaining confidentiality, and making the RETIPS results evident to the organization. Their thoughts on future potential uses for the system range from departmental morbidity and mortality conferences to safety grand rounds, simulation, and lectures.

Carroll and Malmquist make the case that pilots need training to cultivate the foresight and flexibility that are essential to adjust their performance when confronting an unanticipated event. While skilled at following rules for off-nominal conditions, pilots also need interactive and problem-solving skills based on an accurate understanding of their systems. Exposing pilots to low probability failures would make practice possible across the four Resilience Assessment Grid (RAG) potentials: monitor cues to detect anomalies, anticipate implications, use inductive reasoning and problem solving to respond, and learn through debriefing. The authors describe approaches to accomplish this such as Tactical Decision Games (TDG) that require the pilot to make decisions and solve problems while under stress.

Ferreira and Praetorius use Hollnagel's (2012) Functional Resonance Analysis Method (FRAM) to explore the envisioned world of autonomous maritime operations. With data collected from subject matter expert focus groups, the authors used FRAM in multiple scenarios to discover implications for port traffic management under different levels of autonomy. The analyses revealed potential challenges, such as the manner in which shore-based Vessel Traffic Service Centers and Shore Control Centers will need to collaborate to ensure safe operations. They describe the implications for changing from distributed control to polycentric control, the need for improved monitoring and anticipation, interdependencies that can be expected, and critical communication and coordination competencies that maritime traffic management will need as autonomous operations evolve.

Rigaud examines the use of resilience engineering to improve rail organizations' ability to identify and handle variability, whether dealing with regular activity in a challenging environment or with unexpected situations. He shows how workshops, individual interviews, focus groups, and observations can be used to define context, collect data, list factors that need to be either preserved or corrected, and actions that need to be performed. Thorough, organized tables depict how to conduct each of the steps. He concludes by describing a case study to improve scheduling train arrivals and departures.

Cedergren and Hassel examine how resilience engineering can improve the performance of municipal organizations that, while not typically considered a high-risk sector, can be called on to deal with unforeseen challenges. In their three-year effort, they developed an approach for how the city of Malmö, Sweden, could increase the alignment between work-as-imagined and work-as-done. Their research revealed insights about the organization, such as the intentional or semantic structural secrecy that can block collaboration among departments. Their approach sought to improve performance by identifying what each unit does, assess what could quickly lead to negative consequences, identify what activities depend on, learn what backup solutions exist, and illustrate the results. They report how their approach can break patterns of secrecy, improve the potential for accurate judgment, and encourage dialogue.

Lay and Balkin shed light on a field that so far has not been extensively studied, but which poses substantial risks and hazards. Their chapter examines how to improve conditions for utility crews who clear trees from power lines that often pose a mortal threat to safety. They explore the use of resilience engineering to manage risks from being struck by falling limbs to contact with high voltage power lines that are inherent in trimming and removing trees. Their use of the Resilience Assessment Grid (RAG) describes how to anticipate, monitor, respond, and learn from the surprise that is inherent in these highly variable operations. The authors describe a series of methods to improve the resilience of team performance by changing physical or cognitive viewpoint, engaging the unknown, monitoring for weak signals, exploring risk, and reflecting on experience.

2 Methods

Practitioners use a range of methods that have been proven over decades of safety-related research. Alders, Rafferty, and Anderson; Hegde and Jackson; and Ferreira and Praetorious used semi-structured interviews, focus groups, and self-administered surveys. Hegde and Jackson used experiential learning by installing multiple iterations of their RETIPS system to get reactions from users. Bérastégui and Nyssen make the broadest use of methods in their development of a fatigue risk management system, from observation, to brainstorming, semi-structured interviews, focus groups, artefact analysis, simulator studies, workplace trials, and the use of previously validated indicators and measures (e.g., psychomotor vigilance task). Most of the chapter authors used some form of either deductive or inductive analysis. Alders, Rafferty, and Anderson; Carroll and Malmquist; and Lay and Balkin incorporated the Resilience Assessment Grid (RAG), while Praetorius and Ferreira used the Functional Resonance Analysis Method (FRAM) in their exploration of future autonomous maritime traffic management.

3 Themes

Earlier work in resilience engineering often described system adaptive capacity in terms of a response to an unforeseen event, such as a spike in emergency department cases, or natural disaster such as a hurricane. Chapters in this text have turned their attention to themes that may not be as dramatic but are likely to have greater influence long term.

Grounded Understanding Each of the chapters pays careful attention to how work is actually accomplished ("Work-as-Done"). They leverage that understanding to develop solutions that are tailored to their setting they have studied, and that workers can confirm reflects their own lived experience.

Changes to an Existing System to Improve Resilient Performance Most of the chapters use rigorous study to improve current conditions in a range of application areas. Bérastégui and Nyssen develop a more rigorous tailored approach to fatigue assessment to protect operator performance resilience. Sujan uses concepts from resilience engineering to enrich the social dimension of learning in healthcare organizations. Alders, Rafferty, and Anderson use the concepts to analyze and systematically improve healthcare units. Carroll and Malmquist use resilience engineering principles to recommend changes to pilot training so they are exposed to low probability events requiring immediate responses. Cedergren and Hassel look at ways to break down the barriers that limit a municipal departments' ability to collaborate in the face of unforeseen challenges. Lay and Balkin use resilience engineering ideas to cultivate practices among field crews to minimize hazards inherent in tree line clearing, while Hegde and Jackson use them to introduce a collaborative system that enables residents to share their insights into how to improve their work setting. Finally, Rigaud develops an articulated approach that can be applied to improve organization performance, such as rail service scheduling.

Pursuit of an Opportunity Ferreira and Praetorius use a prospective approach to envision possible futures for maritime vessels that are manned by smaller size crews or are fully autonomous, proposing scenarios and using FRAM to examine interdependencies, assess variability, and putting forward thoughts on how to develop systems that perform resiliently.

Each of these themes demonstrates progress in understanding the nature of resilient performance. This advance matters, as technical professionals who are outside of this field need evidence of why resilience engineering concepts should be included in projects, as well as guidance on how it can be done in practice. Managers need proof of why the study of everyday operations deserves resources. In order to meet that need, researchers and practitioners have used criteria and techniques developed in one application sector and shown how they can then be applied to other sectors. They have moved forward in their understanding of how and why systems adapt. They can begin to describe the kinds of systems that are amenable to, or resist, adaptation and have improved our understanding of the implications for resilient performance. Quantitative aspects of performance can be documented and added to qualitative data to become part of the resilience narrative. Such evidence will also address questions about resilient traits (e.g., Haavik et al., 2016), resilience abilities (e.g., DeBoer et al., 2020), and future prospects.

References

Cook, R., & Woods, D. D. (1996). Adapting to new technology in the operating room. *Human Factors, 38*(4), 593–613.

Cook, R., Woods, D. R., & Miller, C. (1998). *A tale of two stories: Contrasting views of patient safety*. National Health Care Safety Council of the National Patient Safety Foundation. American Medical Association.

DeBoer, R., Kaspers, S., Karanikas, N., & Piric, S. (2020). *Measuring safety in aviation: Developing metrics for safety management systems*. Centre for Applied Research Technology, Amsterdam University of Applied Sciences: ISBN: 9789492644206.

Haavik, T. K., Antonsen, S., Rosness, R., & Hale, A. (2016). HRO and RE: A pragmatic perspective. *Safety Science, 117*, 479–489.

Hollnagel, E. (2006). Resilience–The challenge of the unstable. In E. Hollnagel, D. D. Woods, & N. Leveson (Eds.), *Resilience engineering: Concepts and precepts*. Ashgate.

Hollnagel, E. (2011a). Prologue: The scope of resilience engineering. In E. Hollnagel, J. Pariès, D. D. Woods, & J. Wreathall (Eds.), *Resilience engineering in practice: A guidebook*. Ashgate.

Hollnagel, E. (2011b). RAG – The resilience analysis grid. In E. Hollnagel, J. Pariès, D. D. Woods, & J. Wreathall (Eds.), *Resilience engineering in practice. A guidebook*. Ashgate.

Hollnagel, E. (2012a). *FRAM: The functional resonance analysis method for modelling complex sociotechnical systems*. Ashgate.

Hollnagel, E. (2012b). IO, Coagency, intractability, and resilience. In T. Rosendahl & V. Hepsø (Eds.), *Integrated operations in the oil and gas industry: Sustainability and capability development*. IGI Global.

Hollnagel, E. (2014). *Safety-I and safety-II: The past and future of safety management*. Taylor and Francis/CRC Press.

Hollnagel, E., Braithwaite, J., & Wears, R. L. (Eds.). (2019). *Delivering resilient health care*. Routledge.

Hollnagel, E., & Woods, D. D. (2006). Epilogue: Resilience engineering precepts. In E. Hollnagel, D. D. Woods, & N. Leveson (Eds.), *Resilience engineering: Concepts and precepts*. Ashgate.

Hollnagel, E., Woods, D. D., & Leveson, N. (Eds.). (2006). *Resilience engineering: Concepts and precepts*. Aldershot, UK: Ashgate.

Luff, P., Hindmarsh, J., & Heath, C. (2000). *Workplace studies: Recovering work practice and informing system design*. Cambridge University Press.

Nemeth, C., & Herrera, I. (2015). Preface: Building change- Resilience engineering at ten years. In C. Nemeth & I. Herrera (Eds.), *Special issue on resilience engineering* (Reliability engineering and system safety, 141). https://doi.org/10.1016/j.ress.2015.04.006.

Perrow, C. (1984). *Normal accidents: Living with high-risk technologies*. Basic Books.

Wears, R. L., Perry, S. J., & McFauls, S. (2006). "Free fall"– A case study of resilience, its degradation, and recovery in an Emergency Department. In E. Rigaud & E. Hollnagel (Eds.), *Second symposium on resilience engineering*. Juan-les-Pins, France.

Woods, D.D. (2000, October 10). Designing for resilience in the face of change and surprise: Creating safety under pressure. Plenary talk, design for safety workshop, NASA Ames Research Center.

Woods, D. D. (2018). Resilience is a verb: Domains of resilience for complex interconnected systems. IRGC resource guide on resilience. In *Domains of resilience for complex interconnected systems* (Vol. 2, pp. 167–173). EPFL International Risk Governance Center.

Development of Resilience Engineering on Worksites

Akinori Komatsubara

Contents

Many worksite managers are troubled by the aspect of human error on workers. Though they have made various measures such as preparing manuals with guidelines and compelling workers diligently to follow the manuals, it has been impossible to eliminate human error. Field managers are vaguely aware of limitations of those measures that have been previously implemented. Therefore, there are significant expectations from Resilience Engineering, or Safety-II, as a new approach to tackle human error prevention. However, the emphasis that *on-site staff ought to be flexible in adapting to changing worksite conditions* can also lead to problems. For example, we can often hear such remarks from field managers.

"We tell our workers that resilience in the workplace is to adjust to a change of circumstances in your work and to act accordingly. We encourage them to become resilient and flexible workers. Consequently, under the pretext of being very busy with given work, they sometimes cease to follow or violate the manuals that describe the procedures and regulations, which consequently leads to accidents."

"To avoid such accidents, we tell workers to adhere to the manuals once again, and they become confused."

"In other words, they are told, on one hand, to be resilient and flexible, while on the other hand, they are also told to go by the book and stick to the manual."

"How do we explain to our workers to be resilient and flexible while adhering to the manual?"

"How does one describe a resilient worker?"

A. Komatsubara (✉)
Waseda University, Tokyo, Japan
e-mail: komatsubara.ak@waseda.jp

C. P. Nemeth, E. Hollnagel (eds.), *Advancing Resilient Performance*,
https://doi.org/10.1007/978-3-030-74689-6_2

While on-site managers are not necessarily safety professionals, they are generally not academics either. Thus, it is necessary to explain to them—in simple terms—the role and significance of resilience and flexibility, in addition to its relationship with traditional safety approaches and measures.

Therefore, in this chapter, an easy-to-understand scenario will be presented to explain an overview of safety activities, including resilience that must be undertaken at the worksite. Furthermore, the story is intended to draw attention to and emphasize certain key management ideas for resiliency.

1 A Story That Explains Safety

How to Explain Safety-I and Safety-II Whenever an on-site worker plays a certain role in an accident, there is a tendency to use the term *human error or failure*. Bearing this tendency in mind and without avoiding these terms deliberately, the difference between Safety-I and Safety-II could briefly be explained to workers as follows.

Safety-I Achieved only by following predetermined procedures. Deviations from such predetermined procedures are considered human errors. If human errors are avoided, work goals can be successfully achieved with safety. Thus, the goal of Safety-I is to eliminate all human errors. This approach has been from Safety mode of 'centralized control' (Provan et al., 2020), and will focus on adverse event like mishap, failure or accidents (Hollnagel, 2014).

Example: Before engaging in high-voltage electrical repair, the main power supply must be shut off. Performing electrical work without shutting off the power supply is a human error and may lead to accidents. Thus, this error must be eliminated.

Safety-II Be flexible and act according to the current situation (Hollnagel, 2014). If a worker does not have the ability or potential to act according to the situation on hand, accidents may occur. Moreover, even if there were no accidents, in hindsight, there may have been a better way to cope with that situation. In the case, it also could be a failure. Although the outcome was acceptable or even if a successful one for the clients, it is still considered to be a failure from the worker's view. In this sense, the level of success is limitless. Thus, resilience is a means to obtain more desired results; therefore, this will be the approach to focus on success (Hollnagel, 2014, 2018).

Example: Let us consider the case of a doctor performing a medical operation. If the surgery is not performed in accordance with the patient's conditions, it could result in a failure. Even if the surgery is ultimately successful, the doctor may later have regrets upon realizing that there was a better way to operate in which the patient would not have been left with a scar. In this case, the doctor may feel the operation failed. To avoid such a situation, the doctor must build up his or her skills and be more resiliently with careful and flexible behaviour during the operations. It is in this act of refinement of one's potential to read the reality of a situation and make necessary adjustments for the optimal outcome that the doctor becomes a resilient practitioner.

2 Production Activities and the Relationship Between Safety-I and Safety-II

The aforementioned explanations show the differences between Safety-I and Safety-II. However, the relationship between the two, in actual production activities, is still unclear. We present a story to explain this.

Hunting Activity of Primitive Man Let's consider hunting activity of primitive man (Fig. 1). The primary reason for the necessity of safety is to achieve firm production. First, there is a production activity. We wish this activity can be achieved safely. This wish implies the hope that the production staff does not suffer injuries and good-quality service is provided. In the event that safety is the sole concern, production activities ought to be stopped. However, if this were a corporation, then the whole purpose of its existence may be lost.

This situation is no different from that of primitive man. Let us consider primitive man's activity of hunting, wherein the production activity is the capture of prey in good quality condition. If man does not hunt, he will die of starvation. Therefore, production activities cannot be stopped.

Fig. 1 Hunting activity of primitive man

However, the prey is alive, and live prey must be hunted down under changing weather and field conditions.

Efforts at the Individual Level: The Importance of Learning Primitive man may have started out by randomly chasing after his prey. It is likely that he slipped and hurt himself in the pursuit. This is akin to a worker's occupational accident. He may have missed capturing several prey in this manner. This is known as a production mishap. However, there may also have been instances where prey were caught in good quality condition, without any injuries to the man. These are known as successes or the safe fulfilment of production.

Surely, he must have looked back on these accidents and successes and analysed reasons for both. This is known as a root cause analysis. By combining the root cause analysis of failures and successes, primitive man would have learned effective and ineffective (i.e. good and bad) ways of hunting, eventually fine-tuning strategies to locate prey and successfully hunt. These experiences may have been compiled into a guideline manual to share with friends, which could also aid in educating newcomers.

Everyday Learning and Training Learning of guidance is not the only task one must complete before proceeding to hunt. It is equally important to train the muscles. One must have also learned how to effectively use a spear. Technical skills must be built up through everyday learning and training. If all the time is spent on preparation, such as muscle training, and there is not enough time left to hunt, it will be meaningless. In other words, advance preparation that stands in the way of production activities is the evidence of getting one's priorities in reverse order. On the other hand, going out to hunt without proper muscle training is bound to result in accidents or failure. One can only hunt within the range of the strength gained through muscle training activities; hence, to catch big game, it is essential to undergo considerable muscle training. Production activities must be commensurate with the abilities acquired by the individual. In other words, one must acquire the abilities and potentials to meet and match the needs of the production activity.

Going Hunting After sufficient preparation with guidance and physical training, an individual can finally go and hunt, but before hunting, he may anticipate the development of the game of the day and would make a strategy of the game. At the hunting site, hunters must monitor the situation, must be very careful and beware of counterattacks by prey. Situational awareness is indispensable because the prey are 'live'. At precise moments of attacking to respond the prey that appeared, during the hunt, hunters need to make effective and immediate decisions; otherwise, the prey may escape. Thus, non-technical skills, such as monitoring, situational awareness and decision making, are also important attributes (Flin et al., 2008).

Effort at the Team Level: Team Efforts While aiming for big prey that cannot be tackled alone, primitive man must have organised a team to hunt. The size of the team must match the size of the game.

Once the team was organized and before setting out, there must have been a briefing about strategy and the division of roles. The rules for calling out during the course of the hunt would have been set in advance, and hunters would have responded to each other in accordance with the rules. Such communication while chasing the prey cannot be lengthy and must be kept simple. If a strategy was not formed and each person was allowed to act resiliently, without communication, there would be no coordination, and the prey would escape; this may call a functional resonance accident (Hollnagel, 2012a, 2012b). Upon noticing any suspicious movements of the prey—such as an attempt to strike back—the others must be immediately notified by making a loud call. This is known as assertion.

The relationship among team members is also important. If team members are not mutually considerate, one plus one will result in zero. To prevent this, it is necessary to build a good rapport among team members.

An amicable agreement regarding rules for sharing the catch must also be made in advance to avoid future discord among team members. Indeed, it is possible that one team member may experience an injury due to an unforeseeable circumstance despite efforts while hunting. Perhaps he may have received an extra share, notwithstanding the rules. Thus, there must be room to reconsider and revise decisions in an emergency or in an unexpected exceptional situation. Such treatment decisions, actions, and outcomes will accumulate and eventually be incorporated into manuals as precedents, but the precedent is for reference, and it is possible to be changed in the future, in some cases.

As part of routine activities before hunting, the entire team may have prepared by engaging in improvements of the hunting ground, for example, by cutting the grass so their feet do not get caught and to enable them to easily spot prey. This is known as kaizen or workshop betterment. The preparation requires a leader to provide instructions; however, no one will listen to the leader if the leader is a liar or inexperienced person. Thus, good leadership is essential.

Development of Tools With the realisation that working with bare hands has its limitations, primitive man developed spears and bows; this is also known as tool development. They may have improved the tools ergonomically to increase their utility and usability. These improved tools would help those who do not have excellent hunting ability. Then, the rules for correct usage of these tools must also have been established; the blade of a spear must not be grasped. Its handle must be grasped. Grasping the blade will surely cause injury. In other words, technical rules are those that must be followed at all times. Perhaps they compiled the rules into a manual, shared it with newcomers, and, with the manual, trained them on how to use the tools.

With time, they must have thought of ideas to hunt in smaller groups, or perhaps not hunt at all; they may have come up with the idea of digging pits to trap prey. This would have been a labour-saving effort and a move towards automation.

Efforts at the Organizational Level The tribes that worked hard on the aforementioned issues would have hunted well, that is, they would have safely accomplished, produced, and prospered with a high number of prey. The tribes that did not make any effort to tackle these issues would have been unable to catch any prey and declined due to casualties, resulting in an organizational accident. Ultimately, safety is an integral part of production and can be considered as a means to achieve successful production. In this sense, safety measures are important as the very basis of existence for tribes.

Even the tribes that prospered could not afford to be complacent. The population of prey may have gradually declined at the hunting field. In such a case, it may have been necessary for people to migrate. With the introduction of crops, they may have felt that it was better to switch production activities to agriculture rather than rely on hunting. In other words, as natural and social environments around the tribe or organization slowly change, it is important that they pay close attention to signs of change and encourage the community and organization to accordingly adjust; organizational resilience is important, of which potential can be assessed with Resilience Assessment Grid (RAG) (Hollnagel, 2018). This movement may be attributed to organizational culture, especially on flexibility.

3 Lessons That Should Be Learned from This Story

Activities to Ensure Safety Although the entire aforementioned story is fictitious, it is likely an agreeable one.

The story's lesson is that safety is a means to achieve production success. Furthermore, the following list of specific activities could be learned in addition to ways to achieve production safely.

1. Engage in labour-saving and automation activities.
2. Develop tools for production and improve them ergonomically for utility and usability.
3. Conduct improvement or *kaizen* activities for the production site.
4. Secure the number of people necessary for that production.
5. Compile manuals for guidance, rules and regulations. Make these available to everyone. Moreover, make everyone familiarise themselves with these rules and regulations.
6. Improve each person's job capabilities and potentials, including technical and non-technical skills.
7. As part of the production activities of the team, develop non-technical skills for teamwork, such as communication skills, assertion and leadership qualities.
8. The organisation must be aware of changes in the natural and social environment in which it is placed and subsequently direct the organisation in a way that can adjust to the changes. To do so, however, organizational culture on flexibility may be required.

These may be considered to be safety activities to ensure satisfactory production. Furthermore, based on the definitions of Safety-I and Safety-II, steps 1–5 are mainly Safety-I activities and steps 5–8 are Safety-II-based activities.

The Order of Actions Taken for Safety The order in which actions are performed at the site is also important: In other words, there exists a safety management process that includes Safety-I and Safety-II (Komatsubara, 2011). From the eight items listed, steps 1–7, which are directly related to site safety, must be addressed in this order.

For example, let us consider the case of driving a car. If the road is strewn with stones and rubble, it must be cleared (step 3; see Fig. 2). That is the activity to be undertaken first. Attempting to skilfully navigate a rubble-filled road in the name of resilience (step 6) would be a meaningless exercise. Likewise, it is important to establish traffic laws and be orderly when following them (step 5). Instead, trying to drive resiliently in a chaotic traffic situation (step 6) to achieve safety would be counterproductive.

It is also important to offer behaviour assistance facilities to workers (step 2) before training their resilience potentials (step 6). As Fig. 3 illustrates, if we supply rear-view monitors and navigation systems, we must provide them, especially if drivers and workers are novices and do not have such resilience potential.

In other words, in case of a production site, if measures of Safety I can be executed, they must be undertaken as a priority.

Fig. 2 If possible, making the road condition better to prevent needless resilience

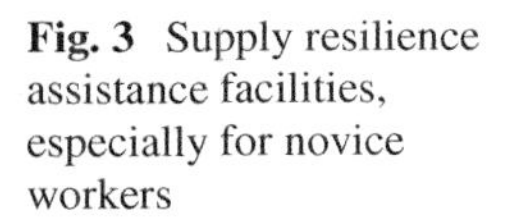

Fig. 3 Supply resilience assistance facilities, especially for novice workers

4 Questions That Arise

Manual and Resilience The relationship between Safety-I and Safety-II as safety activities has been explained by the story in the previous section. Now, we will focus on the relationship between resilience and adhering to the manual. This was the field manager's problem at the onset.

If a manual can be reasonably created, it must be created, and everyone should act in accordance with it. The question that arises is, 'is it good to merely follow the manual?' Or, 'is it good to be slightly flexible with regard to following the manual's rules to adjust to the situation on hand and act resiliently?'

Case 1 illustrates how an issue was avoided by not acting according to the manual. In this case, if the manual instructions had been strictly followed, it would have resulted in utter chaos.

Case 1: Retreating from a Nuclear Power Plant, 2011 It is not formally reported but at the Great Eastern Japan Earthquake (in 2011), a nuclear power plant was said to have allowed workers from inside the plant building to escape outside without measuring their radiation doses. If individual dosimetry had been performed—as prescribed by the manual—the evacuation would have been delayed detrimentally, threatening people's safety.

On the other hand, there are also instances of accidents that have occurred due to the violation of rules in the manual, as in cases 2 and 3.

Case 2: Tokaimura Nuclear Accident in Japan, JCO Plant, 1999 Three workers violated the authorised procedure to produce a small batch of liquid-type uranium fuel, although they were given the procedure. They used an incorrectly sized tank to mix uranium powder into liquid acid, in an attempt to reduce workload and production time. Then, therefore, a criticality accident occurred.

To answer the manager's aforementioned predicament, let us consider the relationship between Safety-I, Safety-II, and the manual.

Types of Manuals In the example of the primitive hunters, there were guides, rules, and regulations, such as manuals. These can be categorised as three types, according to stringency to compliance (Komatsubara, 2016), as follows:

Type 1: Technical regulations – No room allowed for resilient behaviour: a technical procedure, with a physical or natural science background.

For example, when preparing dilute sulphuric acid, concentrated sulphuric acid must be added slowly into the water. If this procedure is reversed, bumping will occur due to the heat of hydration, resulting in the occurrence of an accident. This is similar to the spear usage rule in the case of primitive man: The blade must not be grasped. This procedure is for Safety-I and must not be treated as resilience. The prescribed procedure must be followed under all circumstances. No matter how busy workers are, they should not behave resiliently. Workers are not permitted to change technical procedures for the sake of physical reasons. Established procedures must be followed, and any deviation from the procedure is considered an entirely inadmissible human error.

Type 2: Rules – Resilient behaviour is not acceptable under normal circumstances but it is allowed in emergency situations: a predetermined promise of organisations.

Rules are procedures with a sociological background. They are social promises. Traffic laws are simple examples. Vehicles must keep to the left of the road in Japan and the United Kingdom, and to the right in European countries and the United States. This system chosen by the respective societies is a predetermined arrangement resulting in the smooth flow of traffic, thereby avoiding accidents. Under normal driving conditions, traffic rules must be followed. However, in an emergency, being resilient and deviating from the rule may help achieve better results. In this case, it is permissible to adjust by taking the concept of necessity for an emergency in a legal sense. In the case of primitive man, the amicable agreement for sharing the catch is in this vein; an extra share for an injured member would be acceptable, notwithstanding the agreement.

Type 3: Guides – Resilient behaviour can be accepted, or rather, it is recommended: standard practices that serve as references.

Guides are the manual that defines standard treatment methods. The manual, that defines the methods of support and service to customers in a store, is an example. The manual is a guide and can also be considered a textbook. In the case of primitive man, guidance for chasing and hunting prey would fall into this category. Although this is a set standard, for example, store staffs are expected to change their service approach in a resilient manner depending on the nature of the customer or service targets, situation, and need at the time.

Type 1 requires strict adherence to the procedure. The JCO criticality accident was caused due to the violation of procedures (Komatsubara, 2000). Type 2 requires compliance, to the extent that the rule is a precondition for the procedure. However, in the event of an emergency, it is important to be resilient and take circumstance-appropriate actions rather than following procedures. It is helpful to have an emergency manual, but it is difficult to foresee every type of emergency situation. Hence, an emergency manual shall be a Type 3 manual. In Type 3, the manual serves only as a guide and reference. Achieving good results or success requires good resilience.

This is summarized in Fig. 4. Managers should understand the differences between the three types of manuals based on this figure and explain it to the workers.

Resilient behaviour in types 2 and 3 is not unconditionally allowed for every worker. It is determined by the relative relationship between the magnitude of the situation change and the worker's resilience potential. In short, if the worker's resilience potential is small in relation to the magnitude of the situation change, resilient behaviour is likely to result in undesirable outcomes. It would be beyond his capacity. In order to obtain good outcomes from resilient behaviour, the worker's resilience potential must be large enough for the magnitude of the change.

Figures 5 and 6 are models showing this relationship. Line A indicates the level of resilience potential of the worker. Line B indicates the level of resilience

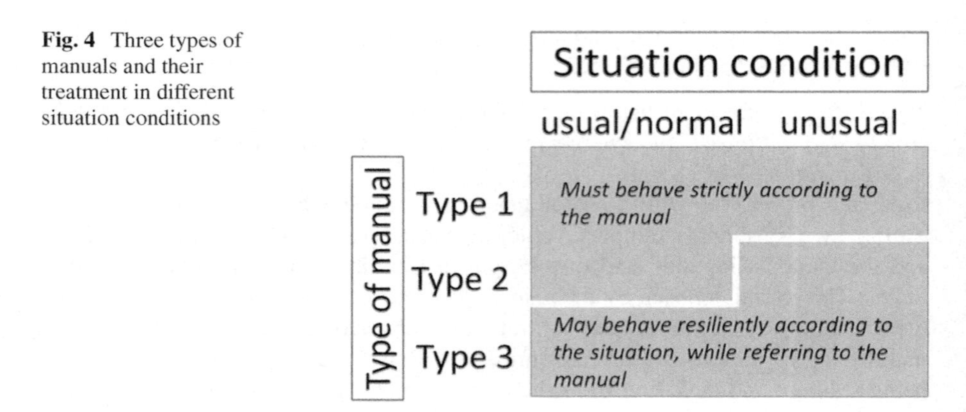

Fig. 4 Three types of manuals and their treatment in different situation conditions

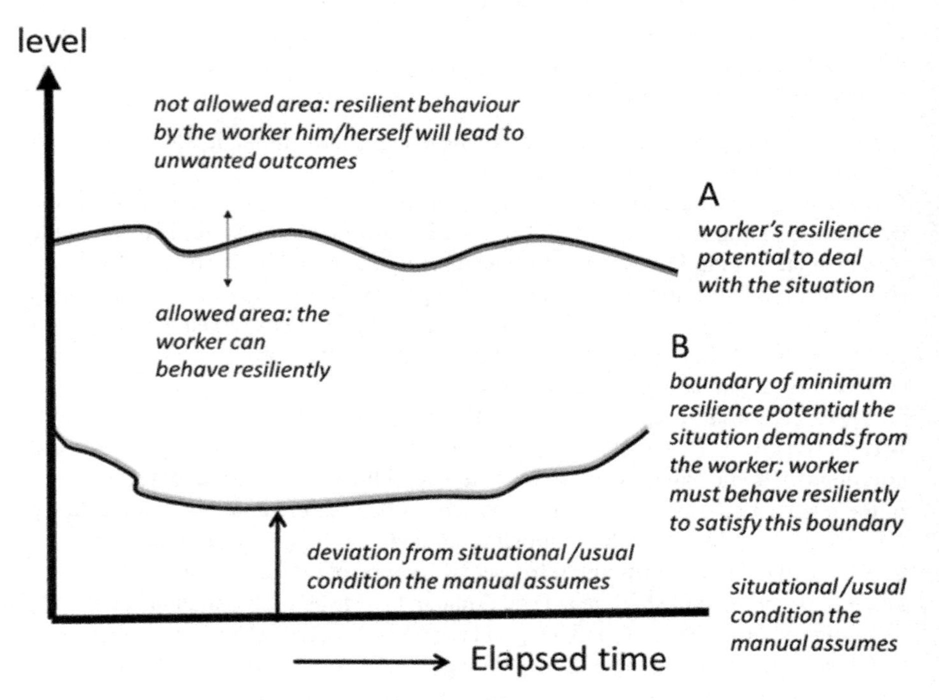

Fig. 5 Relation between resilience potential of the worker and situational demands; case of resilient behaviour being allowed

potential required by the worker at that time. Both lines are wavy, indicating their respective dynamics.

If the worker has a rich resilience potential, the line A moves up. Workers are allowed resilient behaviour in the situation where line B is below line A. On the other hand, line B moves upward as the situation deviates from the normal. As a result, line B goes up beyond line A, resulting in the case presented in Fig. 6. In this

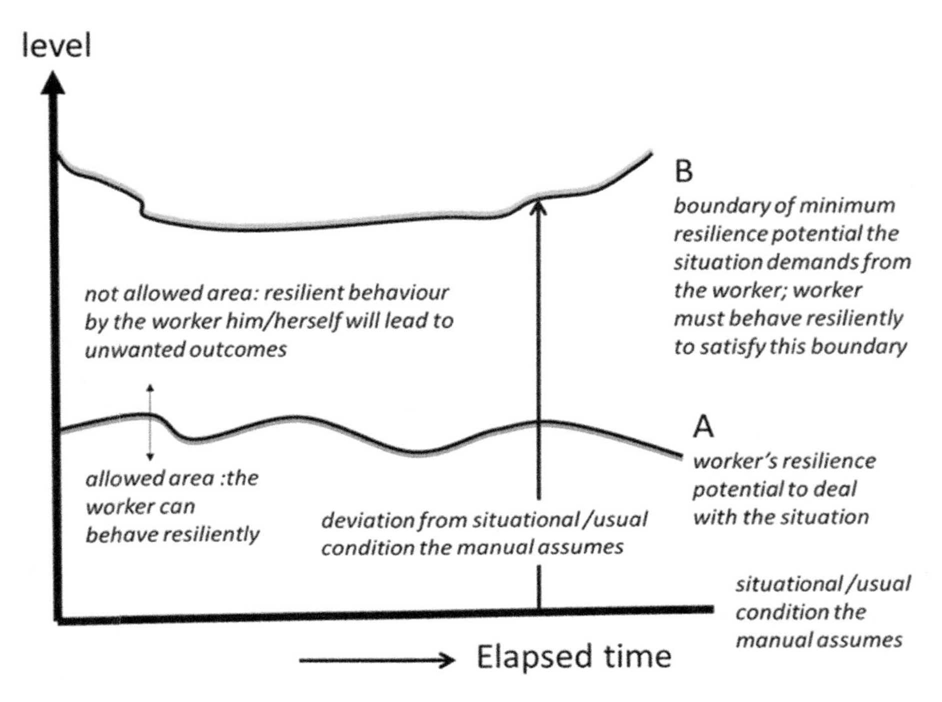

Fig. 6 Relation between resilience potential of the worker and situational demands; case of resilience behaviour not being allowed

case, since the resilience potential of the worker does not meet the demand of the situation, an undesirable outcome shall be obtained if the worker behaves resiliently. Except for real emergencies in which the worker has no choice but to respond himself, the worker should ask to be replaced by another worker with a higher resilience potential or ask his supervisor to give appropriate instructions for coping with the situation.

When line B is far beyond line A, that is, when the situation is far beyond the worker's resilience potential, he will perhaps not be willing to behave resiliently. However, in a real emergency, workers may behave resiliently if no one else can deal with the situation, even if they realize that it is beyond their ability. At such a time, regardless of the outcome, the resilient behaviour would be called a heroic act. However, in the worksite, problems arise when the line B slightly exceeds line A. In this situation, workers often behave with poor resilience potential, saying it is 'probably OK'. This is what the ETTO principles state (Hollnagel, 2012a, 2012b). As a result, accidents often occur. The manager must always tell workers that 'probably OK' is 'never OK'; it is not permitted. When the worker leans towards 'probably OK' resilience, he must first seek the permission of managers with rich potential.

Therefore, the on-site manager must confirm the 'type' each manual belongs to. Fieldworkers must also be clearly informed of the 'type of manual' they are being provided. Type 1 is a manual for Safety-I. Type 2 is a manual for Safety-I under normal circumstances or in the premise of the manual that is determined. However, we can say that Type 3 is a manual for Safety-II.

5 People Prefer Shortcuts

In general, people prefer to use shortcuts. Figure 7 demonstrates this tendency; while the correct route is to walk on the paved section of the road, many people violate this procedure. People generally employ shortcuts in order to save on workload, time and money to achieve their production goals. This behaviour may be instinctive. Therefore, when simply told to follow manuals, workers break the manual due to a tendency towards shortcuts. This is a major issue, especially with Type 1 manuals. This is because the accident will most likely occur, as seen in the JCO criticality accident in case 2. To prevent such accidents, managers need to actively manage workers. Figure 8 summarizes the management strategies that should be implemented. First, low-workload procedures must be constructed; then, the worker must unquestioningly follow the procedures specified in the manual. Since people prefer a lower workload, the shortcut route should be set as the correct procedure, as shown in Fig. 7.

For technical reasons, it may be necessary to define high workload procedures. However, this stimulates shortcut tendencies. To avoid the use of a shortcut, a barrier should be created. In Fig. 3, one of the measures mentioned is to build a strong

Fig. 7 People prefer a shortcut to save on workload, time and money (author's own photo)

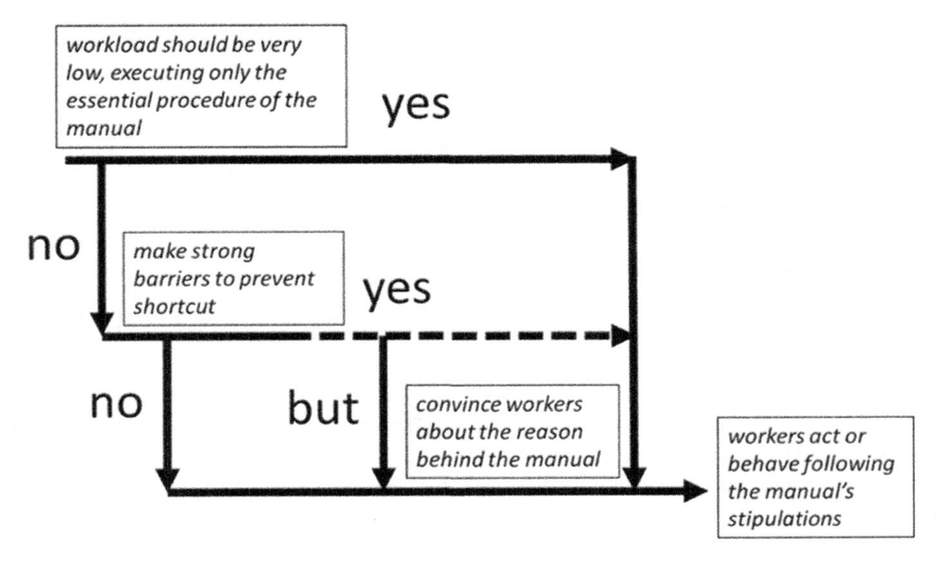

Fig. 8 Management strategies that should be undertaken for persuading workers to follow manuals

fence. However, even if a physical fence is created, it may be destroyed to create a shortcut. To avoid this, managers must explain and convince workers about the reasons for the procedure and make them consciously follow the correct procedure. Even in the case of the JCO criticality accident, if the workers recognized that the reason for the troublesome procedure was to avoid a serious criticality phenomenon, there would have been no resilient violation (Komatsubara, 2000).

The AIDA model—Attention or Awareness, Interest, Desire and Action—old but still frequently used in marketing and advertising may be useful for convincing the workers and calling Attention and Interest to the reasons behind the manual. This stimulates the desire to understand. It will be helpful to explain the danger when the manual is not followed and thus convince them of the reasons for following it. Through this, it is expected that the workers will perform the Actions specified in the manual.

Resilience is Closely Knit to Safety-I The manual should be described at three levels: process, activity and operation or motion. It must also be noted that different types of manuals may exist at different levels within the same job. For example, as aforementioned, to prepare to dilute sulphuric acid, concentrated sulphuric acid must be slowly added to water. This procedure corresponds to Type 1. However, pouring water 'slowly' is a resilient action, and therefore, this part of the procedure calls for Type 3. In other words, Safety-I and Safety-II often exist tightly coupled in the same job. The field manager must let workers understand this fact.

6 Becoming a Resilient Person

The larger a person's resilience potential, the greater their ability to adjust to changes on site; Resilience success cannot be obtained beyond the resilience potential they have. On the subject of manuals, it is possible to significantly deviate from a Type 3 manual and achieve remarkable success. It could even lead to the creation of a more helpful Type 3 manual or guide. A person with high resilience may be called a professional.

Being professional is not merely confined to being knowledgeable. It is impossible to learn all situational responses in advance. Instead, the potential to create answers on the spot, according to the situation, must be cultivated.

Unno Kuniaki (1999) shows a model of skilled technicians, as illustrated in Fig. 9. As skill level increases, the individual becomes more professional and may be considered to be a resilient worker.

Learning is important; however, gaining experience alone does not guarantee improvement in a person's quality as a professional. Moreover, simply learning from production success may sometimes lead to inappropriate practices from safety view, as described in the following example.

Case 3: Explosion of Paint Spray Paint oil tends to harden in the winter, making it difficult to use. Therefore, a senior worker warmed a spray can using hot water in an electric kettle to soften the paint. A younger worker noticed this, and warmed a spray can using the same method, however, without giving any thought to the reason. He warmed it for so long that the can exploded.

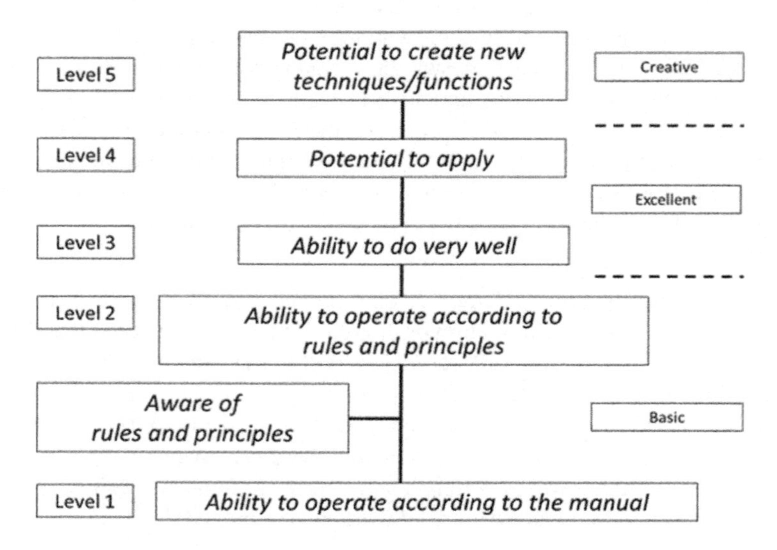

Fig. 9 Skill levels (Adapted from Unno, 1999)

Awareness of reasons about how things work well or work poorly is important. Deep understanding of fundamental principles is essential for good resilience. Indeed, it is important to follow manuals and the guidance they contain, and to learn from successful cases. However, learning the reason behind those successes and, a deep study of principles behind the manual is also necessary. It is only by this process that the desired resilient behaviour can be acquired. The field manager must tell workers not to confuse art with gimmicks; any good results that have been obtained on the surface, if resultant of tricks, any learning should be deterred from it.

7 Conclusion

Resilience is absolutely necessary at every worksite, and Safety-II is indispensable, because a dynamic changing to a greater or lesser extent always occurs. However, to achieve production safety, both Safety-I and Safety-II are imperative. An overall understanding and explanation of safety are needed for all workers. In areas where Safety-I is applicable, activities of Safety-I must first be conducted.

Workers must understand that there are three types of manuals regarding Safety-I and Safety-II. It may be best explained to workers that the concept of a resilient person is a professional who has the potential to be capable of taking actions based on basic rules and principles.

References

Flin, R., O'Connor, P., & Crichton, M. (2008). *Safety at the sharp end: A guide to non-technical skills*. CRC Press.

Hollnagel, E. (2012a). *FRAM: The functional resonance analysis method: Modelling complex socio-technical systems*. CRC Press.

Hollnagel, E. (2012b). *The ETTO principle: Efficiency-thoroughness trade-off: Why things that go right sometimes go wrong*. Ashgate.

Hollnagel, E. (2014). *Safety-I and Safety-II*. Routledge.

Hollnagel, E. (2018). *Safety-II in practice*. Routledge.

Komatsubara, A. (2000). The JCO accident caused by Japanese culture. *Cognition, Technology & Work, 2*(4), 224–226.

Komatsubara, A. (2011). Resilience management system and development of resilience capability on site-workers. *Proceedings of the Fourth Resilience Engineering Symposium*, 148–154.

Komatsubara, A. (2016). *Human factors for safety: Its theory and techniques* (in Japanese), Maruzen Publishing.

Kuniaki, U. (1999). *The succession of the highly expert skills to the next generation* (in Japanese), AGUNE-Syouhu-Sya Publishing.

Provan, D. J., Woods, D. D., Dekker, S. W. A., & Rae, A. J. (2020). Safety II professionals: How resilience engineering can transform safety practice. *Reliability Engineering & System Safety, 195*, open access article 106740.

Fatigue Risk Management System as a Practical Approach to Improve Resilience in 24/7 Operations

Pierre Bérastégui and Anne-Sophie Nyssen

Contents

A growing body of literature indicates that schedules involving extended shifts, night work or other forms of atypical working hours substantially increase workers' fatigue (Chellappa et al., 2019; Doghramji et al., 2018; Czeisler, 2015). These schedules are associated with reduced work performance (Caruso, 2014) and higher risk of errors and accidents (Salminen, 2016; Wirtz, 2010). Despite alarming figures, extended shifts and night work are becoming more common in our so-called 24/7 society. It is estimated that approximately 25% of American workers operate shifts that are not during the daytime (NHLBI, 2005), and nearly 30% work 10 h or more each day (NSF, 2008).

Traditionally, workplace fatigue is almost exclusively managed through limits on the maximum number of hours worked and the minimum duration of rest periods. Governments around the world have imposed a range of legal hours of work limits in attempt to mitigate fatigue-related risk. However, by controlling the amount of worked hours within a specific period, the system does not manage fatigue as a risk factor. Rather, it regulates one – among many others – parameters conditioning operators' fatigue levels. A single-layer normative approach represents a somewhat monolithic view of safety whereby being inside the limits is safe while being

P. Bérastégui (✉) · A.-S. Nyssen
University of Liège, Liège, Belgium
e-mail: pberastegui@etui.org

C. P. Nemeth, E. Hollnagel (eds.), *Advancing Resilient Performance*,
https://doi.org/10.1007/978-3-030-74689-6_3

outside is unsafe. It fails to take into account operational differences and the variability of real-world situations that are likely to affect safety. Forcing a system to adopt norms and practices that proved to be useful in another setting is not only naïve but could actually lead to an increased degradation of the system (Hollnagel et al., 2006). In this context, fatigue risk management systems (FRMS) emerged as a more comprehensive *and pragmatic* approach to mitigate the detrimental effect of fatigue on safety (Dawson et al., 2012). In contrast to traditional prescriptive approaches, FRMS shift the locus of responsibility for safety away from the regulator towards organizations (Gander et al., 2011).

A FRMS can be defined as "a scientifically-based, data-driven addition or alternative to prescriptive hours of work limitations which manages employee fatigue in a flexible manner appropriate to the level of risk exposure and the nature of the operation" (Brown, 2006). Moving away from the traditional hours-of-service restrictions, FRMS propose guidelines on harvesting, developing, implementing and monitoring tentative procedures directed toward fatigue-related risk. The main strength of a FRMS resides in its ecological approach of harvesting and assessing informal strategies currently used within the work group. In that sense, FRMS can be seen as a concrete way to engineer resilience by reintroducing safety managed by humans in addition to safety managed by regulations (Cabon et al., 2011). *Following the principles of resilience engineering (RE), the objective is* to improve the ability of a work system to adjust its functioning during or following disturbances of operators' alertness level in order to sustain required operations under optimum safety conditions. True to the Safety-II approach, FRMS are not confined to the elimination of hazards and the prevention of malfunctions but also aim to continuously improve an organization's potentials for resilient performance – namely "the way it responds, monitors, learns and anticipates" (Hollnagel, 2017a, 2017b).

FRMS rely on two kinds of strategies to ensure organizational resilience in the face of fatigue-related risk. In accordance with the Swiss Cheese Model (Reason, 2000), these strategies are conceptualized as successive defence layers acting at different levels of the potential hazard trajectory (Fig. 1). Fatigue reduction strategies (FRS) aim to reduce the likelihood a fatigued individual is operating in the workplace. FRS can be achieved through the prescription of maximum shift and

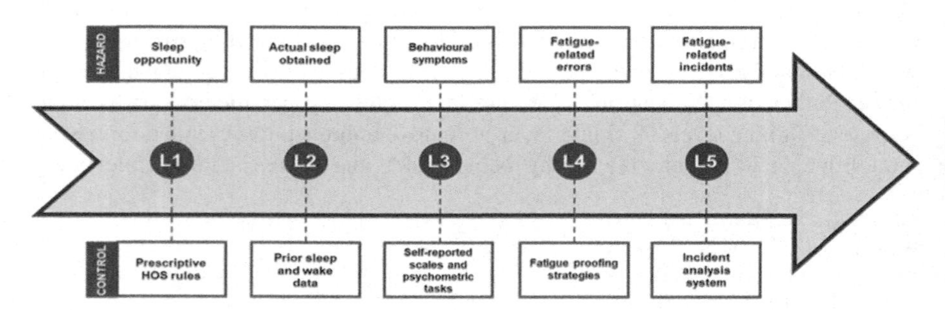

Fig. 1 Fatigue-related risk trajectory with identifiable hazards and controls. (Adapted from Dawson & McCulloch, 2005)

minimum break duration (level 1), the systematic control of sleep hours (level 2), or other behavioural indicators (level 3). In contrast, fatigue proofing strategies (FPS) aim to reduce the likelihood a fatigued individual operating in the workplace will make an error (level 4). FRS and FPS are complementary approaches that must be integrated into a comprehensive FRMS in order to effectively mitigate the level of fatigue-related impairment and its potential consequences (Gander et al., 2017). If a fatigue-related incident occurs despite these four defence layers, level 5 provides an incident analysis framework allowing the organization to improve the effectiveness of level 1–4 and prevent future incidents.

Traditionally, most formal controls addressing fatigue-related risk rely solely on FRS through hours of service regulations (level 1) and do not encompass the notion of fatigue proofing. Interestingly, though, it has been demonstrated that FPS develop as informal work practices in contexts where it is not possible or desirable to further reduce work hours (Bérastégui et al., 2018). The way these informal strategies are developed and consolidated within the workgroup are disorganized, instinctive and unintended. Most of the time, they are observed and passed on through long-standing workplace customs and undocumented mentoring systems (Dawson et al., 2012). Although they emerge as adaptive mechanisms, these individual endeavours may prove to be counterproductive or hazardous. Recently, it has been suggested that the recurrent use of informal FPS may represent a significant risk for the operator in the long run (Bérastégui et al., 2020b). More specifically, the long-term effect of sustained compensatory effort is a draining of workers resources eventually resulting in a breakdown. In this context, the benefits of informal FPS in terms of sustainable performance need to be analysed in relation to the associated costs for the operator. Moreover, individual endeavours participate to widening the gap between work-as-imagined (WAI) by analysts and policymakers, and work-as-done (WAD) by frontline operators (Hollnagel, 2017a). The misalignment of WAI and WAD can make organizations more brittle, as those responsible for managing the work are unaware of the performance adjustments deployed on the job (Sujan et al., 2016). Thus, it is a challenge for fatigue-related risk management to create mutually positive awareness between managers and practitioners in order to reduce this gap and identify counterproductive or harmful strategies. In this context, FRMS proved to be a relevant framework allowing the identification, assessment and formalization of informal strategies (Bérastégui, 2019). Taking advantage of the dynamic nature of WAD (Hollnagel, 2014), FRMS allows resilient performance through a deep understanding of the adjustments that workers undertake on a daily basis.

1 Toward Quantifying Metrics for *Engineering Resilience*

Dawson et al. (2012) outline four main phases for engineering resilience to fatigue-related risk (Fig. 2).

The first phase is to harvest candidate strategies currently used within the work group. The goal is to glean as much information as possible on how fatigue-related

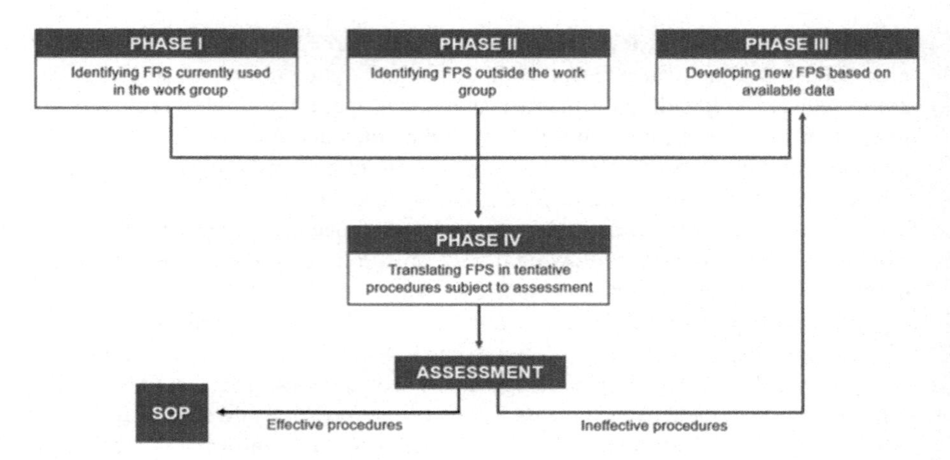

Fig. 2 Phases in the development of a FRMS

risk is handled during day-to-day operations. Field-based qualitative methods such as focus groups or semi-structured interviews are applied to elicit knowledge regarding informal fatigue management. Discussions should revolve around a set of prepared questions to ensure a reasonable level of domain-specific knowledge. Mind mapping adds significant value for generating and structuring ideas during focus groups (Bérastégui et al., 2018). Similarly, visualising or brainstorming specific events may cue additional information during the elicitation process. As a general rule, participants should be encouraged to illustrate their statements with specific events they experienced or witnessed. Ideally, the qualitative knowledge-eliciting techniques should be complemented by a series of parallel workplace observations in order to contextualise the examples communicated during discussions. In cases where an ethnographical focus is unfeasible, supplementary knowledge-eliciting techniques such as open-ended questionnaires can be employed. The end result of this phase will be a comprehensive list of informal FRS and FPS mobilized at the local level. If some of these informal strategies appear to be dysfunctional, countermeasures can be deployed as a matter of priority before moving on to the next step.

The second phase aims to extend strategy identification to similar groups of employees operating outside the work group. The previously described techniques are also suitable here. In addition to discussions with frontline operators, it is valuable to include a subject-matter expert and consult senior managers. Their inputs should shed lights on the organizational specificities likely to hinder the translation of elicited strategies. It may reveal specific professional boundaries and norms in relation to fatigue-risk management, as well as organizational factors or incentives likely to play a role in the integration of standardised strategies into SOP.

During phase 3, the investigators will be looking at developing new strategies based on currently available data. Relevant datasets include records of incidents, near-misses or dangerous occurrences. If necessary, data collection may be expanded using the eliciting techniques described in phase 1. Hazards are then grouped according to thematic areas and prioritised. Selected priority areas are subject to a more in-depth qualitative investigation to determine appropriate countermeasure strategies. Discussions focus on ways to flag the level of elevated risk, to increase levels of error scrutiny or to mitigate error's consequences. Focus groups should include employees at different levels throughout the organization in order to gain a full range of engagement and experience with error management. Information derived from these discussions will be used to support the development of new strategies, adding up to the result of the two preceding phases.

The last phase aims to translate informal strategies harvested during previous phases into tentative procedures subject to assessment. Only procedures that demonstrate clear empirical underpinnings will be integrated into standard operating procedures. Assessment will also provide justification for the deletion of informal strategies at the local level when they are proved to be ineffective or counterproductive. Tentative procedures should be tested individually in order to allow a better understanding of their contribution to the outcomes under study. However, in certain circumstances, it may be more appropriate to evaluate them in clusters based on thematic or technical considerations.

Dawson et al. (2012) propose two distinct assessment approaches. Simulator studies are particularly relevant in settings where observing workers is unfeasible or impractical. It is, however, prone to certain bias making results questionable from an ecological validity standpoint. Participants may exhibit stereotypical behaviours that would not be observed in real-life settings (Peabody et al., 2000). They may be overly watchful due to the expectation of an imminent significant event or exhibit nonchalant attitudes during the exercise due to the absence of real stakes (Datta et al., 2012). Therefore, when possible, a more ecological approach that considers real-life performance should be favoured. Workplace trials have the advantage of being less prone to ecological validity bias but at the expense of a lower degree of control over testing conditions. The main limitation of this approach resides in the difficulty to control for risk exposure. Some external factors are likely to undermine safety in one of the two groups, thus compromising the comparison. Typical cofounders that should be accounted are the number of workers, the number of hours worked, and the proportion of night shifts for each group. Ideally, this approach involves a longitudinal cluster randomised design where workgroups or sites are allocated to experimental (tentative procedures integrated to SOP) or control (SOP only) conditions. The relative performance of the two groups is then compared on the basis of various safety variables (e.g. incident rates, near-misses). If sample size is too small, the allocation to experimental and control conditions is likely to undermine statistical power. In this case, it is preferable to consider procedures' frequency of use as a continuous variable and measure it across all workers. Safety variables are then correlated to identify effective and counterproductive procedures (see Bérastégui et al., 2020b for further details).

Irrespective of which assessment method is put in place, accurate measurement of a wide array of safety performance indicators (SPI) is of paramount importance. There are three types of SPI that should be taken into account for determining procedures' effectiveness.

First are fatigue-related indicators and refer to the first three levels of control of the FRMS (Dawson & McCulloch, 2005). It includes performance tasks such as the Psychomotor Vigilance Task (Basner & Dinges, 2011), and self-reported scales such as the Samn-Perelli Fatigue Scale (Samn & Perelli, 1982). Performance tasks should be favoured since it has been demonstrated that self-reported measures may not always accurately reflect actual fatigue-related impairments (Bérastégui et al., 2020a). The 5-min version of the Psychomotor Vigilance Task is both convenient and sensitive to changes in alertness occurring during extended working hours. However, there may be moments when it is impractical to ask employees to take 5 min to complete a neurobehavioral task. In these circumstances, the use of a single-item subjective rating may be relevant. Other common fatigue-related SPI are sleep-wake histories and can be collected using actigraphy or sleep diaries. Actigraphy is a highly reliable method for objective sleep monitoring with minimal inconvenience to the wearer (Signal et al., 2005). Sleep diaries, on the other hand, are used to collect subjective data on sleep and duty times. They are easy to implement, inexpensive but may show some variability in their accuracy (Gander et al., 2017). Combining the objective data from actigraphy with the subjective data from sleep diaries provides the most accurate assessment of actual sleep-wake history (Girschik et al., 2011).

The second type of SPI are duty-related indictors and include near-misses, errors, incident rates and overall performance (level 4 and 5). These indicators can be collected using self-reporting systems, behavioural checklists or outcome-based approaches. Duty-related SPI are intrinsically linked to the specificities of the operational setting. For data collection to be effective, they should be simple to gather and easy to report. Most importantly, investigators must promote a no-blame culture reflecting an open, trusting and learning atmosphere where everyone can speak about safety issues. Employees participating in the assessment should be assured that no individual information will be shared with colleagues or management. Data collection will preserve anonymity, and analyses will only be conducted to compare and benchmark procedures from a group-level perspective.

Finally, the third type of SPI that should be taken into account relates to employee's quality of work life. Common metrics directly available to the organization are absenteeism, turnover and grievance rates, and tools include the Leiden Quality of Work Questionnaire (van der Doef & Maes, 1999), the Occupational Stress Inventory-Revised (Hicks et al., 2010) and the Maslach Burnout Inventory (Maslach et al., 2016). These metrics are only relevant for long-term workplace trial since they require a certain degree of latency. For shorter trials or punctual simulation sessions, tools such as the NASA Task Load Index (Hart & Staveland, 1988) should be favoured.

2 Aggregating the Data

The success of a FRMS requires the integration of these measurements into a coherent whole, striking a balance between a focus on system safety and employee's quality of work life. This section outlines possible approaches to process the data as well as some of the critical factors that should be considered for data analysis.

It is important to determine the appropriate statistical procedure before starting the investigation. This will determine the size of the required sample and the nature of the conclusions that may be drawn from the results. In cases where the assessment design implies a longitudinal follow-up of employees, statistical analyses have to control for intraindividual correlations (the degree to which repeated measurements for the same participant are correlated). Confounding inter- and intraindividual variability would have enormous consequences for the generalization of the findings. PROC MIXED in the SAS or SPSS software package allows to distinguish the two. Moreover, the use of random coefficients allows for the generalizability of these estimates beyond the particular data sample (IOM, 2004). If the assessment involves only one data point per variable (cross-sectional design), simpler modelling approaches can be employed, such as linear regression for normally distributed data, and Kendall–Theil regression when data are not normally distributed. In all cases, conducted analyses will aim to test the significance of differences between the two groups (control vs experimental) for the variables considered (Fig. 3).

Tentative procedures derived from FRS are assessed based on fatigue-related SPI (level 1–3). It is considered inadvisable to make conclusions based on a single measure of functional status (Gander et al., 2017). Procedures' assessment should involve the widest array of fatigue-related SPI as possible in order to ascertain the validity and accuracy of findings. The hypothesis under study (H_1) is that participants in the experimental condition (implementation of tentative procedures) show significantly lower levels of fatigue than the control condition (SOP only). Typical confounders accounted for include age, drugs intake and sleep history.

Tentative procedures derived from FPS are assessed based on duty-related SPI (level 4–5). The hypothesis under study is that the experimental condition is significantly safer than the control condition (H_2). As described earlier in this chapter, risk exposure differences between conditions should be controlled for. Typical confounders include operator's level of fatigue as well as the number of workers, hours worked and the proportion of night shifts for each group.

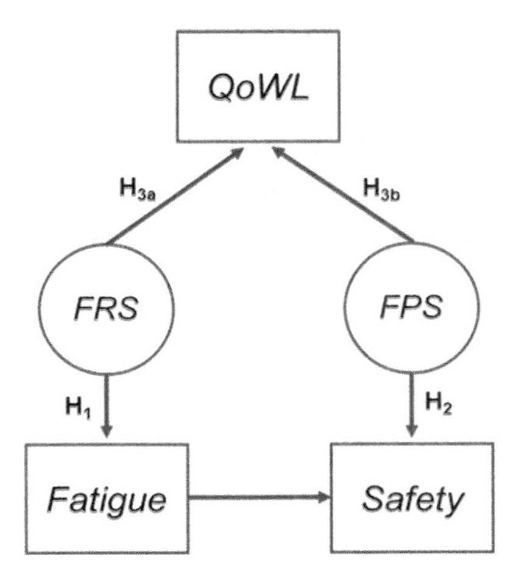

Fig. 3 Tentative procedures assessment

Additionally, both types of tentative procedures should be evaluated from of a quality of work life standpoint (H_{3a} and H_{3b}). The idea is to ensure that, beyond their operational efficiency, these new procedures are not contributing to create an unfavourable work environment for employees (Nyssen & Bérastégui, 2017).

Once the assessment comes to a conclusion, a last round of focus groups may be organized in order to discuss potential optimizations for dysfunctional or unsatisfactory procedures. Reworked procedures should then be subject to a new assessment phase, and so on, until they meet the organization safety standards.

3 Follow-Up and Continuous Improvement

The core principle of a FRMS is to establish a closed-loop process of safety management involving the continuous monitoring of fatigue-related risks and an ongoing development of mitigation procedures. In preceding sections, we outlined the steps for its initial implementation as well as a set of guidelines for data collection and analysis. This final section describe a few key factors that should be considered in follow-up interventions.

Besides developing tentative procedures, a comprehensive FRMS should also pursue its efforts to guarantee their successful implementation in the workplace. The challenge is to disseminate and generalize the new set of procedures to the entire workforce. To this end, procedures should be turned into training materials and integrated into formal education programmes. Employees' learning achievements should be closely monitored to ensure new procedures are properly mastered. Additionally, awareness programmes on fatigue could provide additional support for employees. The objective is to ensure that employees receive regular training on the physiologic consequences of fatigue and learn strategies for maintaining a good sleep hygiene.

Employees should also be given the opportunity to report dysfunctions or failures in the application of procedures. These situations will be thoroughly reviewed in order to identify possible room for improvement. The different control levels described earlier can be employed to strengthen the longitudinal follow-up and tweaking of procedures. Again, due emphasis must be placed on the non-punitive nature of self-reporting. It is crucial to establish an open reporting culture where failures or incidents are considered as learning opportunities rather than faulty behaviours. Otherwise, it is unlikely that an employee will self-identify as fatigued or voluntarily provide information related to a fatigue-related error. Concealing such information could result in a failure to implement new procedures correctly and may potentially pose a greater risk to safety.

4 Example of Implementation

FRMS first appeared in the aviation industry as an alternative approach to the 'one-size-fits-all' model of Flight Time Limitations (FTL). Over the past decades, regulatory authorities gradually allowed airline companies to engineer their own schemes based on an assessment of the conditions that create fatigue in a specific setting. The effectiveness of these initiatives has been demonstrated through a steady decline in the percentage of pilots reporting duty-related fatigue between 1993 and 2006 (ICAO, 2015). Despite these promising results, there have been very few attempts to implement FRMS outside of the aviation industry. This approach could greatly benefit other sectors where fatigue is a significant safety issue.

In a recent study, we deployed a FRMS in the Emergency Department (ED) of a tertiary-care centre in Belgium (Bérastégui, 2019). Emergency physicians (EP) are particularly vulnerable to fatigue due to inconsistent shift rotation, extended duty periods and overnight calls. Following the methodology described in this chapter, we harvested and assessed fatigue management strategies for further integration in SOP.

First, we conducted four focus groups with a total of 25 EP in order to identify strategies deployed to manage fatigue-related risk. EP were asked to describe how on-the-job fatigue affected their efficiency at work and to report any strategies they use to cope with these effects. Using inductive qualitative content analysis, we revealed content themes for fatigue management strategies. Strategies aiming to reduce the subjective experience of fatigue were categorised as FRS, while strategies aiming to mitigate the impact of fatigue on work performance were labelled as FPS. The next step was to assess the efficiency of these strategies. Given the small size of the sample, we opted for a single group design. Each reported strategy was converted in a behavioural item and integrated in a questionnaire assessing frequency of use. We collected fatigue-related SPI using the Psychomotor Vigilance Task (Basner & Dinges, 2011) and the Karolinska Sleepiness Scale (Åkerstedt & Gillberg, 1990). Duty-related SPI were derived from the self-assessment component of the Physician Achievement Review (Hall et al., 1999), and quality of work life SPI consisted of the Maslach Burnout Inventory (Maslach et al., 2016). All instruments were combined into a practical and functional Android-based application installed on a smartphone device. Each physician was briefed on when and how to report each type of SPI on the smartphone. Analyses were conducted to determine the association between SPI and strategies' frequency of use. By doing so, we were able to identify effective strategies and dysfunctional ones.

We were able to identify 12 FRS and 21 FPS (see Bérastégui et al., 2018 for details). FRS mainly consisted of rest-time management, physical exercise and food or energy drink intake. FPS were comprised of self-regulation, task-reallocation and error-monitoring strategies. For instance, EP working night shifts tend to complete patient records as and when it comes rather than letting things pile up to compensate for the impact of fatigue on short-term memory. Similarly, physicians reported deferring complex but not urgent tasks during the night shift to colleagues working the following morning. Other examples included double-checking for tasks regarded as 'vulnerable' to fatigue-related risk, and verbalizing acts or prescriptions to avoid omission.

Assessment revealed that the use of FRS was associated with decreased levels of fatigue while preserving satisfactory levels of quality of work life. Similarly, FPS allowed EP to sustain adequate work performance despite sleep deprivation. However, the analysis of quality of work life revealed that some FPS represent a significant risk for EP's well-being over the longer term. Specifically, scores on the emotional exhaustion sub-dimension of MBI were found to be positively associated with FPS frequency of use. Besides demonstrating the feasibility of applying this methodology in emergency care, our findings also stress the importance of considering quality of work life SPI as some strategies resulted in a trade-off between work efficiency and quality of working life. It allowed the ED to identify these dysfunctional strategies and engage a reflection on potential countermeasures. Effective strategies, on the other hand, were considered for implementation in SOP. The identification of at-risk operators, task redistribution within the team, or increasing standard checks for at-risk operators are examples processes that are still, at the time of writing, under further scrutiny (Bérastégui, 2019).

Other sectors may greatly benefit from the implementation of FRMS. This is especially the case of the ride-hailing industry that has grown exponentially in recent years. The sector faces unparalleled transformations due to the emergence of the so-called gig economy, transforming into a fee-for-service, unregulated taxi industry. With this transformation comes two key regulatory and safety challenges that deserve attention. First, most drivers are employed in a primary job and work in the ride-hailing industry during their time off. Cumulating multiple jobs is likely to lead to extended periods of wakefulness or during nights – two factors that increase the risk of driving accidents. Second, drivers are employed as independent contractors and, in this respect, are not obliged to undergo a medical examination. This poses a significant risk for safety as medical problems such as obstructive sleep apnoea are associated with reduced levels of alertness. In face of these challenges, the gig economy mostly promotes the 'internalisation of external risks' (Holts, 2018) by shifting most of the risk of doing business from the company to individual gig workers. This general trend toward self-management strengthens the economic model of platform work at the expense of hidden human costs. Recently, the American Academy of Sleep Medicine (AASM) stated that fatigue and sleepiness are inherent safety risks in the ride-hailing industry and urged companies and regulatory authorities to work together to address this public safety issue. According to the AASM, this collaborative effort should be in the form of FRMS and more stringent regulations (Berneking et al., 2019). Applying this framework would allow to fully grasp the scope of this issue in the gig economy and to engineer countermeasures tailored to the specificities of platform work. Moreover, mobile applications used by ride-hailing companies offer many possibilities for collecting SPI in a timely and systematic manner. However, the primary obstacle remains the lack of incentives or enforcement measures for platform companies to take responsibility of risk management.

5 Conclusion

The ongoing development of 24/7 operations in various industries stress the need of a more tailored and comprehensive approach to manage fatigue-related risk. The main limitation of the traditional prescriptive approach is that it does not take into account the specific conditions that creates fatigue in a given environment. Moreover, it overlooks the importance of preserving margin of manoeuvre in complex adaptive systems. Organizations with insufficient margin of manoeuvre are likely to fall into maladaptive traps leading to systems failures (Woods Branlat, 2011). With FRMS, organizations move from the illusion that fatigue-related risk can be managed through one-size-fits-all prescriptive measures, and develop procedures tailored to the specificities of the work environment. By relying on a wide range of means and resources, FRMS enable more robust safety management than the single defensive layer of prescriptive regulations (Gander et al., 2017).

In line with the RE perspective, it encapsulates a broader focus than identification of safety hazards only. Specifically, it acts on the four abilities that are necessary for a system to be resilient (Hollnagel, 2011):

- *Knowing what to look for* (what is or can become a threat), through constant monitoring of relevant SPI
- *Knowing what to expect* (how to anticipate threats and opportunities), through successive defence layers acting at different levels of the potential hazard trajectory
- *Knowing what to do* (how to respond to disturbances), through the development, assessment and implementation of effective countermeasures
- *Knowing what has happened* (how to learn from experience), through an incident analysis framework aiming to prevent future fatigue-related incidents

In that sense, a FRMS is about how resilience can be engineered in the context of fatigue-related risk through concrete measures acting on each of these four factors. Moreover, by building on current hours of service regulations, it combines both regulated and managed safety – two notions that are regarded as complementary from a RE standpoint (Falzon, 2014). It relies on all of the available resources, namely, the existing rules and standards enacted by regulatory authorities, and the ad hoc procedures constructed locally to cope with the variability of real-world situations. Such approach is often described as 'adaptive safety' (Falzon, 2011) as it relies on the intelligence of the agents involved in everyday activities. The FRMS literature is laying great emphasis on the fact that employee's expertise can provide critical insights regarding safety issues. This view is comforted by several studies showing that mitigation strategies develop as informal work practices when they are not addressed at the organizational level (Bérastégui et al., 2018; Schulte et al., 2015; Dawson et al., 2012). Similarly, RE research has demonstrated the value of performance variability of frontline practitioners to deal with uncertainty (Sujan et al., 2015; Nyssen & Blavier, 2013). Variability in everyday performance is the reason why things go right as it ensures a certain degree of system flexibility in response to varying conditions. True to Safety-II, FRMS are moving from a 'quick fix' philosophy and introduce a closed-loop process of safety management through the ongoing development of second-order solutions. It explores the ways in which workers have the potential to be flexible when systems may not have been perfectly designed or when conditions are challenging. The insights gained by such naturalistic approaches allow a deep understanding of the causal dynamics at stake (Sheps & Wears, 2019) in a manner conducive to learning and system improvement.

It may be tempting to conclude that the spontaneous development of informal practices demonstrates the underlying capacity of the work system to self-regulate. However, especially in occupations associated with a high level of motivation and commitment, this can lead to pushing individual resources to their limits and losing all margins of manoeuvre. In this case, resilience at the organization level solely relies on resilience of individuals, at the expense of a draining of resources eventually resulting in a breakdown (Bérastégui et al., 2020b). Moreover, individual endeavours may represent a significant risk for the overall organization in the long term resulting from the misalignment of WAI and WAD (Hollnagel, 2017a). Thus, it is the responsibility of the organization to support the development of formal procedures and to provide employees with appropriate resources to keep pace with work demands. Otherwise, the lack of formal procedures will shift the strain to the employees' own resources to sustain safety, causing a subsequent risk of depletion. It is our belief that moving the 'burden' of adaptation from the individual to the system is a key element in achieving resilient performance in 24/7 operations, and that the FRMS framework provide a concrete approach to do so.

References

Åkerstedt, T., & Gillberg, M. (1990). Subjective and objective sleepiness in the active individual. *International Journal of Neuroscience, 52*(1–2), 29–37.

Basner, M., & Dinges, D. F. (2011). Maximizing sensitivity of the Psychomotor Vigilance Test (PVT) to sleep loss. *Sleep, 34*(5), 581–591.

Bérastégui, P. (2019). *La gestion du risque associé à la fatigue en médecine d'urgence: Identification et évaluation de pratiques informelles*. [Doctoral thesis, University of Liège, Belgium]. https://orbi.uliege.be/handle/2268/236178

Bérastégui, P., Jaspar, M., Ghuysen, A., & Nyssen, A. S. (2018). Fatigue-related risk management in the emergency department: A focus-group study. *Internal and Emergency Medicine, 13*(8), 1273–1281. https://doi.org/10.1007/s11739-018-1873-3.

Bérastégui, P., Jaspar, M., Ghuysen, A., & Nyssen, A. S. (2020a). Fatigue-related risk perception among emergency physicians working extended shifts. *Applied Ergonomics*. https://doi.org/10.1016/j.apergo.2019.102914.

Bérastégui, P., Jaspar, M., Ghuysen, A., & Nyssen, A. S. (2020b). Informal fatigue-related risk management in the Emergency Department: A trade-off between doing well and feeling well. *Safety Science, 122*, 104508. https://doi.org/10.1016/j.ssci.2019.104508.

Berneking, M., Rosen, I. M., Kirsch, D. B., et al. (2019). The risk of fatigue and sleepiness in the ridesharing industry: An American Academy of Sleep Medicine position statement. *Journal of Clinical Sleep Medicine, 14*(4), 683–685.

Brown, D. (2006). Managing fatigue risk: Are duty hours the key to optimising crew performance and alertness? *Proceedings of the Flight International Crew Management Conference*, Brussels, Belgium.

Cabon, P., Deharvengt, S., Berechet, I., Grau, J.-Y., Maille, N. P., & Mollard, R. (2011). From flight time fatigue risk management systems- A way toward resilience. In E. Hollnagel, J. Pariès, D. D. Woods, & J. Wreathall (Eds.), *RE in practice-A guidebook* (pp. 69–86). Ashgate.
Caruso, C. C. (2014). Negative impacts of shift work and long work hours. *Rehabilitation Nursing, 39*(1), 16–25.
Chellappa, S. L., Morris, C. J., & Scheer, F. A. J. J. L. (2019). Effects of circadian misalignment on cognition in chronic shift workers. *Scientific Reports, 9*(1), 699. https://doi.org/10.1038/s41598-018-36762-w.
Czeisler, C. A. (2015). Duration, timing and quality of sleep are each vital for health, performance and safety. *Journal of the National Sleep Foundation, 1*(1), 5–8.
Datta, R., Upadhyay, K., & Jaideep, C. (2012). Simulation and its role in medical education. *Medicine Journal of Armed Forces India 68*(2), 167–172. https://doi.org/10.1016/S0377-1237(12)60040–9
Dawson, D., & McCulloch, K. (2005). Managing fatigue: It's about sleep. *Sleep Medicine Reviews, 9*(5), 365e80.
Dawson, D., Chapman, J., & Thomas, M. J. (2012). Fatigue-proofing: A new approach to reducing fatigue-related risk using the principles of error management. *Sleep Medicine Reviews, 16*(2), 167–175.
Doghramji, K., Tanielian, M., Certa, K., & Zhan, T. (2018). Severity, prevalence, predictors, and rate of identification of insomnia symptoms in a sample of hospitalized psychiatric patients. *The Journal of Nervous and Mental Disease, 206*(10), 765–769. https://doi.org/10.1097/NMD.0000000000000888.
Falzon, P. (2011). Rule-based safety vs adaptive safety: An articulation issue. *3rd International Conference on Health Care Systems, Ergonomics and Patient Safety (HEPS)*. Oviedo, Spain.
Falzon, P. (2014). *Constructive ergonomics*. CRC Press.
Gander, H. L., Powell, D., Cabon, P., Hitchcock, E., Mills, A., & Popkin, S. (2011). Fatigue risk management: Organizational factors at the regulatory and industry/company level. *Accident; Analysis and Prevention, 43*(2), 573–590.
Gander, P. H., Lora, J. W., van den Berg, M., Lamp, A., Hoeg, L., & Belenky, G. (2017). Fatigue risk management systems. In M. Kryger, T. Roth, & W. C. Dement (Eds.), *Principles and practices of sleep medicine* (pp. 697–707). Elsevier.
Girschik, J., Fritschi, L., Heyworth, J., & Waters, F. (2011). Validation of self-reported sleep against actigraphy. *Journal of Epidemiology, 22*, 462–468.
Hall, W., Violato, C., Lewkonia, R., Lockyer, J., Fidler, H., Toews, J., Jennett, P., Donoff, M., Moores, D. (1999). Assessment of physician performance in Alberta: the physician achievement review. *CMAJ 161*(1), 52–57.
Hart, S., & Staveland, L. (1988). Development of NASA-TLX (Task Load Index): Results of empirical and theoretical research. In P. Hancock & N. Meshkati (Eds.), *Human mental workload* (pp. 139–183). North Holland.
Hicks, R. E., Bahr, M., & Fujiwara, D. (2010). The occupational stress inventory-revised: Confirmatory factor analysis of the original inter-correlation data set and model. *Personality and Individual Differences, 48*(3), 351–353.
Hollnagel, E. (2011). Prologue: The scope of resilience engineering. In E. Hollnagel, J. Pariès, D. Woods, & J. Wreathall (Eds.), *Resilience Engineering in practice–A guidebook* (pp. xxix–xxxix). Ashgate.
Hollnagel, E. (2014). *Safety-I and Safety-II: The past and future of safety management*. Ashgate.
Hollnagel, E. (2017a). RE and the future of safety management. In N. Möller, S. O. Hansson, J. E. Holmberg, & C. Rollenhagen (Eds.), *Handbook of safety principles* (pp. 25–41). Wiley.
Hollnagel, E. (2017b). *Safety-II in practice: Developing the resilience potentials*. Routledge.
Hollnagel, E., Woods, D. D., & Leveson, N. (2006). *Resilience engineering. Concepts and precepts*. Ashgate.

Holts, K. (2018). *Understanding virtual work: Prospects for Estonia in the digital economy*. https://www.riigikogu.ee/wpcms/wp-content/uploads/2017/09/Virtual-work-size-and-trends_final1.pdf

Institute of Medicine. (2004). The study of individual differences: Statistical approaches to inter- and intraindividual variability. In IOM (Ed.), *Monitoring metabolic status:Predicting decrements in physiological and cognitive performance*. Committee on Metabolic Monitoring for Military Field Applications. National Academies Press.

International Civil Aviation Organization. (2015). *Fatigue management guide for airline operators*. (2nd ed.). https://www.icao.int/safety/fatiguemanagement/FRMS%20Tools/FMG%20for%20Airline%20Operators%202nd%20Ed%20(Final)%20EN.pdf

Maslach, C., Jackson, S. E., & Leiter, M. P. (2016). *Maslach burnout inventory: Manual* (4th ed.). Mind Garden.

National Heart, Lung, and Blood Institute. (2005). *Your guide to healthy sleep* (NIH publication No. 06-5271). U.S. Department of Health and Human Services.

National Sleep Foundation. (2008). *Sleep in America poll: Summary of findings*. https://www.sleepfoundation.org/sites/default/files/2018-11/2008_POLL_SOF.pdf

Nyssen, A. S., & Bérastégui, P. (2017). Is system resilience maintained at the expense of individual resilience? In J. Braithwaite, R. L. Wears, & E. Hollnagel (Eds.), *Resilient health care III: Reconciling work-as-imagined and work-as-done* (1st ed., pp. 37–46). CRC Press.

Nyssen, A. S., & Blavier, A. (2013). Investigating expertise, flexibility, and resilience in socio-technical environments: A case study in robotic surgery. In E. Hollnagel, J. Braithwaite, & R. L. Wears (Eds.), *Resilient health care* (pp. 97–110). Ashgate.

Peabody, J., Luck, J., Glassman, P., Dresselhaus, T. R., & Lee, M. (2000). Comparison of vignettes, standardized patients, and chart abstraction: A prospective validation study of 3 methods for measuring quality. *JAMA, 283*(13), 1715–1722. https://doi.org/10.1001/jama.283.13.1715.

Reason, J. (2000). Human error: Models and management. *BMJ, 320*, 768–770.

Salminen, S. (2016). Long working hours and shift work as risk factors for occupational injury. *The Ergonomics Open Journal, 9*(1), 15–26.

Samn, S., & Perelli, L. (1982). *Estimating aircrew fatigue: A technique with implications to airlift operations* (Technical Report No. SAM-TR-82-21) (pp. 1–26). USAF School of Aerospace Medicine.

Schulte, A., Donath, D., & Honecker, F. (2015). Human-system interaction analysis for military pilot activity and mental workload determination. In *Proceedings of 2015 IEEE International Conference on Systems, Man, and Cybernetics* (pp. 1375–1380).

Sheps, S., & Wears, R. L. (2019). 'Practical' resilience: Misapplication of theory? In J. Braithwaite, E. Hollnagel, & S. H. Hunte (Eds.), *Working across boundaries: Resilient health care* (Vol. 5). CRC Press.

Signal, L. T., Gale, J., & Gander, P. (2005). Sleep measurement in flight crew: Comparing actigraphic and subjective estimates to polysomnography. *Aviation, Space, and Environmental Medicine, 76*, 1058–1063.

Sujan, M. A., Chessum, P., Rudd, M., Fitton, L., Inada-Kim, M., Spurgeon, P., & Cooke, M. W. (2015). Emergency care handover (ECHO study) across care boundaries: The need for joint decision making and consideration of psychosocial history. *Emergency Medicine Journal, 32*, 112–118.

Sujan, M. A., Pozzi, S., & Valbonesi, C. (2016). Reporting and learning: From extraordinary to ordinary. Resilient Health Care. In J. Braithwaite, R. L. Wears, & E. Hollnagel (Eds.), *Resilient health care III: Reconciling work-as-imagined and work-as-done* (pp. 37–46). CRC Press.

Van der Doef, M., & Maes, S. (1999). The Leiden Quality of Work Questionnaire: Its construction, factor structure, and psychometric qualities. *Psychological Reports, 3*(Pt 1), 954–962.

Wirtz, A. (2010). Lange arbeitszeiten. untersuchungen zu den gesundheitlichen und sozialen auswirkungen langer arbeitszeiten. *Bundesanstalt für arbeitsschutz und arbeitsmedizin, récupéré*. http://oops.unioldenburg.de/volltexte/2010/996/pdf/wirlan10.pdf, le 07.12.17.

Woods, D. D., & Branlat, M. (2011). Basic patterns in how adaptive systems fail. In: *Resilience engineering in practice* (pp. 127–144). Farnham, UK: Ashgate.

Using the Resilience Assessment Grid to Analyse and Improve Organisational Resilience of a Hospital Ward

Matthew Alders, Anne Marie Rafferty, and Janet E. Anderson

Contents

1 Introduction

Helping organisations to perform in a resilient manner is an emerging area of research in healthcare, but despite philosophical development there remains a lack of practical tools that can be used by practitioners. Tools and methods for analysing resilient performance are needed to inform organisational improvement. This chapter describes a new method for analysing resilience in hospital systems based on the Resilience Assessment Grid (RAG). The RAG is a tool for analysing the four resilience potentials (Hollnagel, 2018) proposed to underpin resilient system performance: anticipating, monitoring, responding, and learning (Hollnagel, 2010). Its purpose is to assist users to analyse their own system and diagnose areas of weakness by answering a series of questions about the four resilience potentials. However, the original questions proposed in the RAG were high level and abstract. They were designed to be adapted to the local context in which it was to be applied and used in a survey administered to staff, but there was little guidance provided for adapting the questions to the context. Previous research has used different methods for

M. Alders (✉) · A. M. Rafferty · J. E. Anderson
King's College London, London, UK
e-mail: matthew.alders@kcl.ac.uk

C. P. Nemeth, E. Hollnagel (eds.), *Advancing Resilient Performance*,
https://doi.org/10.1007/978-3-030-74689-6_4

developing contextual RAG survey items, but these are either too close to the original theoretical items and not very clear for healthcare professionals (Hunte, 2016; Engvall et al., 2017) or include conceptual additions which are not developed sufficiently for understanding resilient performance (Van der Beek & Schraagen, 2015; Rigaud et al., 2015).

The process described in this chapter starts with a detailed exploration of Work-as-Done (WAD) as a basis for developing the questions, rather than starting with the original RAG questions and adapting them. Resilient healthcare theory distinguishes between Work-as-Imagined (WAI) in policies and procedures, and Work-as-Done (WAD) in practice. The variability of demands in a complex sociotechnical system means that it is impossible to plan for all eventualities (Hollnagel, 2010; Hollnagel et al., 2015), or specify all necessary actions in advance. People working in the system must adapt WAI to ensure successful performance across a range of conditions. Adaptive actions to manage evolving demands and problems – Work-as-Done – are, therefore, vital for good performance.

The three stages of the process we developed are:

1. Exploring Work-As-Done through focus groups with the people who understand the work
2. Generating survey items about responding, monitoring, learning and anticipating activities from the focus group data and administering the survey to gather different perspectives on the resilience of the system
3. Reflecting on the survey results and identifying potential interventions to support or improve/strengthen organisational resilience

In this chapter, we will describe how we used this process to analyse the potential for resilient performance of an Acute Medical Unit in a large central London hospital.

2 Background to the RAG

Over the course of its development, the Resilience Engineering (RE) community has proposed a range of approaches for defining, characterising and modelling organisational resilience (Rigaud et al., 2015). A narrative review we conducted showed that the resilience engineering literature for measuring, analysing or evaluating resilient performance is fragmented and inconsistent (Alders, 2019). Many initiatives appeared to be initial ideas proposed in conference proceedings or book chapters and lacked sufficient conceptual and methodological development for application in practice.

The two most well-known approaches for analysing organisational resilience are the Resilience Assessment Grid (formally Resilience Analysis Grid) (RAG) (Hollnagel, 2010) and the Functional Resonance Analysis Method (FRAM) (Hollnagel, 2012), which differ conceptually and methodologically. FRAM proceeds by identifying an area for improvement or analysis and maps the relevant functions and their interactions (Hollnagel, 2016). There is a growing literature documenting experience using FRAM (Clay-Williams et al., 2015; Raben et al., 2018; Ross et al., 2018; O'Hara et al., 2020), but few applications of the RAG. This is possibly due to the need for more methodological development of the RAG, which we hope to address in this study.

The purpose of RAG is to provide a detailed characterisation of organisational operations that can be used to manage and develop an organisation's potential for resilient performance (Hollnagel, 2018). It is important to draw a distinction between resilience as a quality of a system and resilient system performance (Hollnagel et al., 2006; Hollnagel et al., 2015). Studying resilience as a quality of a system is not helpful because it only emerges in response to stressors and, therefore, cannot be measured directly. Instead, analysis should focus on the potential for resilient performance (Hollnagel, 2018) and the activities that enable systems to perform in a resilient manner. The four resilience potentials are theorised to underpin resilient performance and to be equally necessary and jointly sufficient for resilience to emerge (Hollnagel, 2018). They can, therefore, be used as proxy measures for resilient performance. Currently, these are known as the four resilience potentials (Hollnagel, 2018), but previously they have been known as the four cornerstones or the four abilities for resilient system performance (Hollnagel, 2010; Hollnagel, 2015). All these terms refer to the same four concepts:

1. The potential to respond – being able to respond to expected and unexpected disturbances or take advantage of opportunities; by employing prepared responses or by changing the current mode of functioning
2. The potential to monitor – being able to look for factors that may affect the performance of the system in a positive or negative sense.
3. The potential to learn – being able to learn the right lessons from the right experience
4. The potential to anticipate – being able to anticipate developments (both positive and negative) further into the future (Hollnagel, 2018).

The RAG uses diagnostic questions (adapted to the local context) to assess how effectively the four resilience potentials are being performed. The answers can be used to produce a detailed characterisation of each of the four potentials as well as of an organisation's overall potential for resilient performance (Hollnagel, 2018).

3 Rationale for the Approach

Resilient performance is an emergent property of system performance, and the potential for it can be understood by viewing how everyday work is performed (Woods, 2006; Hollnagel et al., 2013; Hollnagel et al., 2015). Resilience is distributed in various ways across system functions and activities (Anderson et al., 2016) and emerges during performance as people, processes and equipment simultaneously interact and react to meet the constantly fluctuating demands of complex systems. Therefore, studying everyday work is a good starting point for understanding resilient performance. We used a social constructivist perspective to inform our research methodology, which means that understanding the social processes and interactions between people working in the system and their environment is important.

Previously, researchers have used a variety of methods for analysing resilient performance, including either qualitative or quantitative methods (Hollnagel, 2010; Mendonça & Wallace, 2015; Jain et al., 2018), but there is growing consensus that mixed methods are an effective way to study complex phenomena (Berg et al., 2018). We modified an explanatory sequential mixed method research design (Creswell & Clark, 2011) to create three distinct (but interconnected) research phases. We used a first phase of qualitative data collection to inform the development of survey items, a second phase of quantitative data collection using the survey items (Creswell & Clark, 2011), which we then used to inform a third (and final) exploratory qualitative phase to identify potential interventions. This resulted in a replicable, systematic method for generating relevant items to analyse the organisational resilience of a complex system at micro, meso and macro levels.

4 Overview of New Method

In the following sections, we will describe each of the three stages in more detail, with the intention that the reader will be able to replicate this approach as a practical guide for use in their own complex system. We used this three-stage process to analyse the potential for resilient performance of an Acute Medical Unit (AMU) in a large, central London teaching hospital. AMUs are inpatient wards specifically staffed and equipped to treat patients presenting with acute medical (distinct from surgical intervention) illness from Emergency Departments (EDs) or the community for assessment, care and treatment (Scott et al., 2009). They have multidisciplinary teams that assess and manage both medical illness and functional disability (Bell et al., 2008). Such units face complex challenges related to the high patient acuity, pressure for beds, the need for close coordination with the Emergency Department and other hospital wards, and the expectation that patients' length of stay will be short before they are referred on or admitted to a ward, typically within a few days.

5 Phase 1: Understanding Work-as-Done

Data Sources The aim of phase one was to achieve a detailed understanding of WAD on the AMU (from the perspective of nursing staff) as a basis for developing survey questions in Phase 2. The people who work in the system have the best understanding of its daily functioning, and so it is important to understand what they experience as challenging as it is likely to be important for understanding resilient performance. We used focus groups with nurses representing the range of roles and experience present on the AMU to understand WAD. Eighteen nurses took part in nine focus groups, with two to five participants in each focus group. To ensure staff engagement in the process, we started by discussing what nurses find challenging about their work. We assumed they are the experts of their system, so what they find difficult to manage is likely to be the 'sticking points' of working in the system. By asking general questions such as 'What is hard about your work?', 'Why is it hard?' and 'How do you manage your challenges?' we started to generate an understanding of the WAD within their system. Crucially, it was important to talk about the challenges of everyday work in the language that nurses used.

Data Analysis We conducted a detailed, theoretically informed thematic analysis of the focus group data. We used a combined deductive-inductive approach. We used the CARe Resilience Model (Anderson et al., 2016) (Fig. 1) as a theoretical lens for the initial deductive analysis.

To understand the focus group data in more detail, we then moved to an inductive analysis, informed by the four resilience potentials. This allowed for a broader and deeper analysis of the focus group data than a simple descriptive analysis could provide.

Results The thematic analysis of the focus group data provided rich insights into the nurses' WAD on the AMU, including the highly complex nature of nurses' everyday clinical work and the importance of social interactions for achieving work goals, including enabling adaptations. Table 1 summarises the main topics elicited from nurses about the challenges of their work.

All these challenges required nurses to anticipate, respond, monitor, and learn. For example, managing patient flow required the ability to anticipate patient and organisational needs, respond to immediate patient and organisational concerns, monitor how the process was being managed to identify delays and other problems, and learn about the best way to manage patient flow.

6 Phase 2: Staff Survey

Data Sources Data from Phase 1 were used in this phase to develop the survey questions. The survey was then administered and analysed.

Data Analysis The thematic analysis completed in Phase 1 was reviewed line by line to generate questions. This resulted in a long list of potential questions that was then reduced by combining questions with similar meanings and removing duplicates. We then reviewed the questions to determine which of the four resilience potentials were represented. Some resilience potentials were not represented well

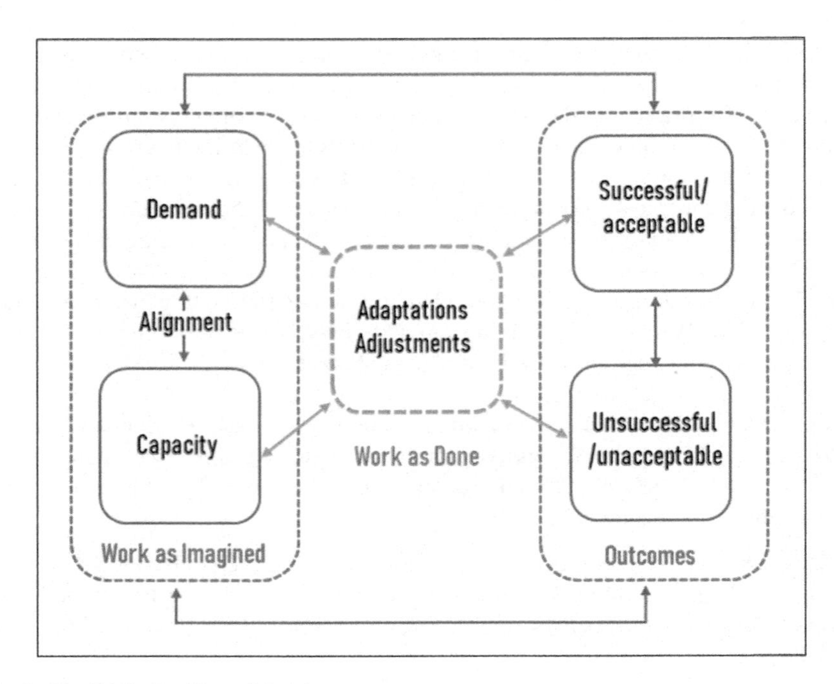

Fig. 1 The CARe Resilience Model

Table 1 Challenges in everyday work identified in focus groups

Challenge	Examples
Coordination with Emergency Department	There was sometimes a mismatch between information conveyed during handover and the patient's condition on arrival to the unit. This meant staff were unprepared and sometimes lacked vital information for patient care.
Deteriorating patients	Responding to patient deterioration and escalating to the medical team involved balancing demands on the wider team and risk to the patient.
Skill mix	Newly qualified nurses and agency staff lacked the expertise of more experienced staff.
High tempo work	Multiple simultaneous admissions and discharges created high workload and the need to coordinate within the unit and with other teams.
Staffing level	Staffing shortages due to sickness and staff turnover were common.
Teamwork	Teamwork was variable and sometimes poor.
Equipment	Equipment shortages and breakdown were common.
Patient flow	Coordinating admissions, transfers and discharges was a challenge and involved balancing competing demands.
Challenging patients	Patients with aggressive behaviour, confusion, substance use disorders or mental health problems required more time and expertise.
Time management	Nurses constantly had to prioritise tasks and patient needs according to the time available.
Patient acuity	Many patients were clinically complex and required specialist expertise that had to be coordinated.

Table 2 Example survey items from the RAG for the AMU

Resilience potential	Example of survey item
Responding	Coordinating with the multidisciplinary team to facilitate the complex discharge of a patient
	Appropriately escalating a deteriorating patient to a senior colleague earlier than the National Early Warning Score (NEWS) recommendation
Monitoring	Knowing when nursing colleagues in your zone need help
	Knowing how busy other zones are compared to yours
Learning	Communicating the learning from things that have gone well, despite challenges
	Changing practice in response to learning from incident reports
Anticipating	Identifying when the workload on the next shift will be high
	Taking action to reduce workload for the next shift

because they were not discussed very much in the focus groups. For example, there were few questions about learning. We, therefore, added questions where there were few data to guide question development. Therefore, the final survey included items generated from the focus group discussions and theoretically generated items. Focus group participants then reviewed the questions and made suggestions, and the questions were edited into a self-administered survey format with 5-point Likert scales. The final survey included 37 items referring to specific system activities such as responding to deteriorating patients, managing security problems, and coordinating admissions and discharges. Participants were asked to rate how well they thought the system activities were conducted on a scale from one to five. Table 2 shows examples of survey items for each of the four resilience potentials.

The survey was administered to all 77 nursing staff on the unit using online and paper methods. Fifty-five completed surveys were available for analysis.

The small number of respondents meant that analysis was restricted to descriptive statistics. In most systems, there will be small numbers of participants available to respond to a survey, which will restrict the level of statistical analysis that can be performed. However, the intention of this process is not to reduce organisational resilience to statistical values but to gather a wide range of perspectives on the most important activities which can then be summarised to inform improvement efforts. Survey data were analysed to provide information on the range of staff who completed the survey and the mean scores for all items, and each of the four resilience potentials to allow for comparison.

Results The response rate was 71%. The survey was completed by 55 staff, including 14 health care assistants, 23 junior nurses, 8 senior nurses and 10 managerial nursing staff. Table 3 shows an overview of the results.

Table 3 Summary of survey results

Resilience potential	Number of items	Mean (SD)	Lowest scoring items
Responding	23	3.84 (.45)	Including mental health nurses in handover to support patients with mental health needs
			Involving nursing team members in assessing a patient's mental capacity
			Changing staff allocation during a shift in response to changed workload
			Agreeing the allocation of tasks between colleagues working in the same bay
Monitoring	6	3.41 (.35)	Knowing how busy other zones are compared to yours
			Coordinating patient transfers from different zones to the same destination
			Informing team members when there are new admissions coming into the zone
Learning	5	3.70 (.84)	Communicating the learning from things that have gone well despite challenges
			Communicating the learning from incident reports
			Changing practice in response to learning from things that have gone well
Anticipating	3	3.79 (.23)	Taking action to reduce workload for the next shift

Results showed that the highest rated activities were responding and the lowest were monitoring. The mean scores were used to identify areas needing further discussion in Phase 3 to identify improvements.

7 Phase 3: Identifying Potential Improvement Interventions

Data Sources Seven focus group participants participated in the final phase of the process. Group semi-structured interviews were used to review and discuss the survey results and consider what system-level interventions could improve the potential for resilient performance. The following topic guide was used to structure the interview:

- Were you surprised by the results?
- Does it support what you already knew?
- Were any system activities rated better or worse than you expected?
- Do you think any of these activities could be improved?
- How do you think these activities could be improved?

Data Analysis A descriptive analysis of the interview data was used to identify opportunities for improvement. A list of suggested interventions was generated and presented to senior staff for further discussion and potential implementation.

Results There was overall agreement with the survey results, with only a few participants expressing surprise about the results for some of the survey items. Most of the discussion focused on the survey items with the highest and lowest mean scores. The nurses provided detailed contextual descriptions that deepened understanding of why some system activities on the AMU were done well and why some were much less effective.

Nursing staff identified interventions for improving each of the resilience potentials, including training, changing the organisation of work and redeploying unused system resources. Table 4 summarises example interventions for each of the four resilience potentials.

Although the suggestions were high level and not specified in detail, they provided a basis for ongoing discussions within the unit about opportunities for improvement. Nurses supported the view that developing system-level interventions could reduce unwanted variability from relying on individuals to fulfil system-level functions and provide support to manage the complexity of their everyday clinical work more effectively. The granularity of the survey items meant that it was relatively straight forward to identify ways to improve the system because they represented specific activities on the unit.

Interventions were submitted to senior staff for consideration and have not yet been implemented.

Discussion The study advanced the RAG by developing a replicable, systematic process for healthcare professionals to analyse the resilience potential of their healthcare systems and identify potential interventions. Although the interventions have not as yet been implemented, focus group data showed that respondents thought the process focused attention on known problems that had not been addressed, such as integration of mental health nurses into the team, and identified new areas for improvement, such as monitoring the distribution of workload. The process systematically documented areas for improvement that could then be prioritised for action.

Concepts from resilient healthcare were successfully used to engage healthcare professionals to identify opportunities for quality improvement, based on an analysis of the organisation's potential for resilient performance. Because the process started with exploration of the challenges of everyday clinical work, it was possible to develop specific survey questions that were relevant to the system being analysed. Each stage of the multi-phase process is built on the analysis of the previous stage, and in an iterative manner each stage is provided a different perspective for understanding system performance. The suggested improvements generated through this process addressed system-level deficiencies and directed attention to these issues.

Table 4 Examples of suggested interventions

Resilience potential	Proposed intervention
Responding	More training to support effective working with mental health nurses
	Improve communication systems between nursing team and psychiatric liaison team
	Involve mental health nurses in handovers
Monitoring	Huddles between nurses in charge and coordinators in each area to review workload and capacity
	Develop clearer criteria for admission to the unit
Learning	Involve junior staff in investigation processes
	Share suggestions for improvement between the unit and the Emergency Department
Anticipating	Enable senior staff to share anticipatory actions with other staff
	Anticipate workload on the next shift and reduce workload by completing more tasks during current shift

This is helpful because many healthcare improvement efforts focus on training and sanctioning individuals and neglect system improvements.

The findings were restricted to the nursing population on the AMU. However, the results showed that much of the complexity of everyday clinical work involved interaction between different healthcare disciplines and organisational units. Cross-boundary interaction is known to be a source of complexity in sociotechnical systems (Woods, 2006; Hollnagel et al., 2015). Given that the study focused on nursing work only, future research could examine the views of the multidisciplinary team about how the system functions and include other organisational units.

The RAG should be an ongoing process and the questions should be repeated over time to analyse how a system improves or changes (Hollnagel, 2010; Hollnagel, 2015). However, there is little direction about how long there should be between repeated applications or what should trigger this repeated application. We considered a number of options for repeated applications of this three-stage process over time. One option could be to repeat the same survey questions after implementing interventions to assess whether the potential for resilient performance has improved. A second option could be to repeat the whole three-stage process to generate new survey items and focus on new responding, monitoring, learning and anticipating activities. Perhaps, a third intermediate option could be to update the survey items using focus groups rather than repeating Phase 1.

There was some imbalance in the number of items for each of the four resilience potentials. The survey results indicated how much the system focused on responding, with a lack of balance between the four resilience potentials. Theoretically, this limited the AMU's potential for resilient performance (Hollnagel, 2010). However, it was unclear what the balance of the four resilience potentials should be. Resilience engineering theory recognises that different systems need different balances between the four resilience potentials (Hollnagel, 2010; Hollnagel, 2018), but it does not provide any suggestion for understanding what the balance needs to be for a given system. This is a question that could be explored in future research.

The survey items developed in this initial application of the three-stage process were specific to the local system demands of the AMU from the perspective of the nursing staff who work there. It is unclear whether the same questions could be used across different settings in the same industry, for example, different inpatient hospital wards, outpatient clinics or even community healthcare settings. Applications of this three-stage process across different healthcare settings would allow for the comparison of different system challenges. It would also allow for comparison between the survey items of different healthcare systems to reach a higher level of understanding about how responding, monitoring, learning and anticipating manifest across different healthcare systems. Perhaps, there will be a core set of questions that are relevant across healthcare systems, with specific questions relevant to features of specific healthcare systems. However, there needs to be more applications of this process to understand this better. To aid the development of this understanding, new questions could be saved in a shared repository so that other researchers and healthcare professionals can see the questions used in other healthcare systems. Finally, formal evaluation of whether the process described here results in interventions that are better defined and targeted, and whether quality is improved as a result is now required.

8 Conclusion

We have described a theoretically informed systematic process for analysing the potential for resilient performance in healthcare that can be replicated in other clinical settings. The process involves first understanding WAD using focus groups with key informants. Focus group data are analysed to identify areas of system activity that are crucial for resilient performance. Survey questions are then developed and administered to the whole workforce to gain a wide range of views on how resilient the system is. The results are analysed to identify opportunities for improvement. The hallmarks of this process are the close engagement with experts to understand work as done and identify improvements. This allows a fine-grained analysis of the resilience in the system that is not possible with questions that are not tailored to the system being analysed.

References

Alders, M. (2019). *A reflective process for analysing organisational resilience to improve the quality of care*. [Doctoral thesis, King's College London]. https://kclpure.kcl.ac.uk/portal/en/theses/a-reflective-process-for-analysing-organisational-resilience-to-improve-the-quality-of-care(446b7cab-6384-4451-a6c4-c1b1794e563c).html

Anderson, J. E., Ross, A., Back, J., Duncan, M., Snell, P., Walsh, K., & Jaye, P. (2016). Implementing resilience engineering for healthcare quality improvement using the CARE model: A feasibility study protocol. *Pilot and Feasibility Studies, 2*(1). https://doi.org/10.1186/s40814-016-0103-x.

Bell, D., Skene, H., Jones, M., & Vaughan, L. (2008). A guide to the acute medical unit. *British Journal of Hospital Medicine, 69*(Supp 7), M107–M109. https://doi.org/10.12968/hmed.2008.69.Sup7.30432.

Berg, S. H., Akerjordet, K., Ekstedt, M., & Aase, K. (2018). Methodological strategies in resilient health care studies: An integrative review. *Safety Science, 110*, 300–312. https://doi.org/10.1016/j.ssci.2018.08.025.

Clay-Williams, R., Hounsgaard, J., & Hollnagel, E. (2015). Where the rubber meets the road: Using FRAM to align work-as-imagined with work-as-done when implementing clinical guidelines. *Implementation Science, 10*(1), 125. https://doi.org/10.1186/s13012-015-0317-y.

Creswell, J., & Clark, V. P. (2011). *Designing and conducting mixed methods research* (2nd ed.). Sage.

Engvall, C., Ekstedt, M., & Ros, A. (2017). To improve the potential for resilience at a pediatric ward. *Proceedings of the 6th Resilient Health Care Network Meeting*. Vancouver.

Hollnagel, E. (2010). Epilogue: RAG—The resilience analysis grid. In E. Hollnagel, J. Paries, D. Woods, & J. Wreathall (Eds.), *Resilience engineering in practice: A guidebook*. Ashgate.

Hollnagel, E. (2012). *FRAM: The functional resonance analysis method: Modelling complex socio-technical systems*. Ashgate.

Hollnagel, E. (2015). *RAG—The resilience analysis grid*. http://erikhollnagel.com/onewebmedia/RAG%20Outline%20V2.pdf

Hollnagel, E. (2016). *A brief introduction to the FRAM*. http://functionalresonance.com/brief-introduction-to-fram/index.html

Hollnagel, E. (2018). *Safety-II in practice: Developing the resilience potentials*. Routledge.

Hollnagel, E., Woods, D., & Leveson, N. (2006). *Resilience engineering: Concepts and precepts*. Ashgate.

Hollnagel, E., Braithwaite, J., & Wears, R. (2013). *Resilient health care*. Ashgate.

Hollnagel, E., Wears, R., & Braithwaite, J. (2015). *From Safety-I to Safety-II: A white paper*. The Resilient Health Care Net. University of Southern Denmark, University of Florida, USA, and Macquarie University, Australia.

Hunte, G. (2016). Engineering resilience in an urban emergency department, part 2. *Proceedings of the 5th resilient health care network meeting*. Middlefart, Denmark.

Jain, P., Mentzer, R., & Mannan, M. S. (2018). Resilience metrics for improved process-risk decision making: Survey, analysis and application. *Safety Science, 108*, 13–28. https://doi.org/10.1016/j.ssci.2018.04.012.

Mendonça, D., & Wallace, W. A. (2015). Factors underlying organizational resilience: The case of electric power restoration in New York City after 11 September 2001. *Reliability Engineering & System Safety, 141*, 83–91. https://doi.org/10.1016/j.ress.2015.03.017.

O'Hara, J., Baxter, R., & Hardicre, N. (2020). 'Handing over to the patient': A FRAM analysis of transitional care combining multiple stakeholder perspectives. *Applied Ergonomics, 85*, 103060. https://doi.org/10.1016/j.apergo.2020.103060.

Raben, D., Viskum, B., Mikkelsen, K. L., Hounsgaard, J., Bogh, S. B., & Hollnagel, E. (2018). Application of a non-linear model to understand healthcare processes: Using the functional resonance analysis method on a case study of the early detection of sepsis. *Reliability Engineering & System Safety, 177*, 1–11. https://doi.org/10.1016/j.ress.2018.04.023.

Rigaud, E., Neveu, C., Langa, S., & Obrist, M. (2015). Sociotechnical system resilience assessment and improvement method. *Proceedings of the 6th resilience engineering association symposium*. Lisbon, Portugal.

Ross, A., Sherriff, A., Kidd, J., Gnich, W., Anderson, J., Deas, L., & Macpherson, L. (2018). A systems approach using the Functional Resonance Analysis Method to support fluoride varnish application for children attending general dental practice. *Applied Ergonomics, 68*, 294–303. https://doi.org/10.1016/j.apergo.2017.12.005.

Scott, I., Vaughan, L., & Bell, D. (2009). Effectiveness of acute medical units in hospitals: A systematic review. *International Journal for Quality in Health Care, 21*(6), 397–407. https://doi.org/10.1093/intqhc/mzp045.

van der Beek, D., & Schraagen, J. M. (2015). ADAPTER: Analysing and developing adaptability and performance in teams to enhance resilience. *Reliability Engineering & System Safety, 141*, 33–44. https://doi.org/10.1016/j.ress.2015.03.019.

Woods, D. (2006). Essential characteristics of resilience. In E. Hollnagel, D. Woods, & N. Leveson (Eds.), *Resilience engineering: Concepts and precepts*. Ashgate Publishing.

Learning from Everyday Work: Making Organisations Safer by Supporting Staff in Sharing Lessons About Their Everyday Trade-offs and Adaptations

Mark Sujan

Contents

We know a lot about the tragic consequences of failures in healthcare. For the past 20 years, there has been much public and media attention on high-profile healthcare incidents with catastrophic outcomes for patients as well as the healthcare professionals involved. Examples come to mind all too easily: the scandal around appalling standards of care at Mid Staffordshire NHS Foundation Trust that resulted in as many as 1200 patients dying needlessly has laid bare systemic failings at this organisation and more widely in the National Health Service (NHS); and the tragic death of 6-year old Jack Adcock at Leicester Royal Infirmary has become synonymous for the excessive pressures and demands that NHS emergency departments place on their staff, many of whom are left without proper supervision and support. These examples paint a dire and sobering picture of specific instances of the poor quality of care provided to patients. Without disregarding the harm and distress that these scandals and incidents have caused, there is another way of looking at this: despite having to work within a notoriously underfunded and overstretched health system, individuals, teams and organisations routinely provide good quality care to millions of patients every day. Aside from dedication and hard work of staff, we know

M. Sujan (✉)
Human Factors Everywhere, Ltd., Woking, UK

C. P. Nemeth, E. Hollnagel (eds.), *Advancing Resilient Performance*,
https://doi.org/10.1007/978-3-030-74689-6_5

surprisingly little about how such daily successes are achieved, how we can learn from these, and how, as a result, care can be improved further.

The mainstream patient safety movement is, arguably, still quite young when compared with other safety-critical industries, such as commercial aviation and the energy and process industries. Patient safety became an important topic for politicians and health policy makers with the publication of the Institute of Medicine report "To err is human" in 1999 in the USA (Kohn et al., 2000), and the subsequent publication of the report "An organisation with a memory" in the UK in 2000 (Department of Health, 2000). These reports used data from earlier studies, such as the now famous Harvard Medical Practice Study (Brennan et al., 1991), and extrapolated these findings to produce figures that captured (and shocked) public imagination: as many as 98,000 patients might die every year in the USA as a result of healthcare errors.

From the start, these reports placed great emphasis on building capacity for organisational learning within health systems. The UK report carries this in its title (*An Organisation with a Memory*), as does the more recent (2013) so-called Berwick report that sets out recommendations for the UK government in response to the findings of the public inquiry into the failings at Mid Staffordshire (National Advisory Group on the Safety of Patients in England, 2013). The Berwick report is called "A promise to learn – a commitment to act", and it suggests that the NHS should aim to become a system devoted to continuous learning and improvement.

With this sustained focus on promoting organisational learning within health systems, one might expect to see significant progress and improvement with patient safety. However, the available evidence suggests differently (Wears & Sutcliffe, 2019). There is now a wealth of literature that demonstrates that healthcare organisations continue to struggle to generate useful learning from past experiences, and that they routinely fail to translate learning into meaningful and sustainable improvements in practice (Kellogg et al., 2017; Macrae, 2015; Peerally et al., 2016). The literature has identified numerous barriers to effective learning from experience in healthcare. Examples include fear of blame and repercussions, poor usability of incident reporting systems, lack of feedback to staff, and lack of visible and sustainable improvements to working practices and the working environment (Anderson et al., 2013; Braithwaite et al., 2010; Sujan, 2015; Tucker & Edmondson, 2003).

The argument in this chapter is that the struggles with organisational learning in healthcare are, at least in part, due to the narrow way in which learning has been cast as learning from incidents (LFI), without proper consideration of how healthcare professionals actually deliver care (the "work-as-done") and how the learning processes need to be embedded and supported within an organisation. This approach to learning only considers the few extraordinary situations, where a system has broken down, that is, organisations are seeing only half the picture at best. A resilience engineering (RE) approach that focuses on learning from everyday work (work-as-done) enables organisations to learn about why, most of the time, things go right, and how the manifold adaptations and trade-offs within a healthcare environment can prevent everyday disturbances and disruptions from turning into catastrophes (Hollnagel et al., 2015).

The next section reviews LFI theory more generally and discusses some of the major shortcomings with the narrow implementation of LFI in healthcare. Then, a short critique of LFI from a RE perspective is given, and an approach to learning from everyday work based on RE thinking is outlined, and its application in a multi-site study is described. The chapter concludes that healthcare organisations should adopt the RE perspective to create a more positive, inclusive, and ultimately more effective learning environment for improving patient safety. The proposed approach is one such way in which organisations can implement a RE approach to organisational learning.

1 Learning from Incidents

The literature and the concepts around organisational learning are very broad, and there is no universally agreed definition (Easterby-Smith et al., 2000). Organisational learning is sometimes described as a continuous cycle of action and reflection, which can take place at different levels, such as individual, group, organisation or even a business sector (Carroll & Edmondson, 2002). In safety-critical industries, an important approach to organisational learning is learning from incidents. Ideally, effective LFI triggers improvements in practice that enhance safety and productivity. The analysis of incidents seeks to reveal contributory factors and underlying causes, which can be addressed in order to reduce the likelihood of incidents recurring.

There is currently a lot of renewed interest in LFI, and the literature on LFI is growing. There has been a collection of papers providing analysis, reflection and critique of LFI in a recent special issue on this topic in the journal *Safety Science* (Stanton et al., 2017). A number of integrative frameworks have been proposed that demonstrate the depth and breadth of LFI (Drupsteen & Hasle, 2014; Jacobsson et al., 2012; Lindberg et al., 2010). These frameworks describe LFI as a process that includes not only the actual investigation of incident data, but also the steps that take place before and after, such as data gathering, identifying improvements, implementation and evaluation.

Lukic and colleagues developed an empirical model for LFI with subsequent extensions and refinements, which emphasises the social and organisational enablers for effective learning rather than the specific steps in the LFI process (Lukic et al., 2012; Lukic et al., 2010). This highlights the fact that learning is a social process and that effort and resource should be dedicated not only to improving the quantity and the quality of the data, but also the social infrastructure for effective learning.

Several papers in the special issue (see above) provide evidence that organisations across different sectors still seem to struggle with getting good LFI processes off the ground (Littlejohn et al., 2017; Margaryan et al., 2017; Rollenhagen et al., 2017). Organisations are often reasonably good at collecting, analysing and disseminating a lot of incident data, but then fail to link this to meaningful learning and changes to practice. In their analysis, Margaryan et al. (Margaryan et al., 2017) very usefully observe that LFI tends to rely on insights from safety science and human factors, but has so far neglected to tap into the body of knowledge around the wider literature on adult workplace learning. LFI processes usually sit within risk management and safety departments, with little input from learning and development experts. As a result, organisations collect a wealth of incident data, but access to data by itself does not guarantee that any learning or changes to practice take place. This requires opportunities for collective sensemaking, deeper reflection ("double-loop learning" in the terminology of Argyris & Schön (Argyris & Schön, 1996)), and proper linking of safety information to professional practice (Stanton et al., 2017).

In healthcare, policy makers looked towards other industries for guidance and lessons about LFI. The end product was the widespread adoption of Root Cause Analysis (RCA) for the investigation of incidents with significant patient harm, and the implementation of organisation-wide and national incident reporting systems. In the NHS, the National Reporting and Learning System (NRLS) was set up in 2003 to collect and aggregate incident reporting data at a national level. NRLS has built up a repository of millions of incident reports, but there is little evidence that this has contributed to any kinds of significant and sustainable improvements in patient safety (Carruthers & Phillip, 2006; Vincent et al., 2008).

Numerous studies of LFI in healthcare have investigated the barriers to reporting and learning, and there appears to be an emerging consensus that in its current shape and form, it is simply not working. As alluded to in the previous section, criticisms that have been raised include inadequate feedback to staff who contribute incident reports, lack of visible improvements to clinical practice, the development of weak improvement interventions focusing largely on staff education and procedure compliance, and the use of LFI as a management rather than improvement approach (Westbrook et al., 2015). In addition, LFI can be perceived as contributing to the existing blame culture, because there is a temptation to focus on what individuals did wrong. The exclusive focus on LFI as a vehicle for organisational learning in healthcare also neglects other, more informal learning mechanisms, such as local communities of practice (Sujan, 2015). The breadth of these criticisms has prompted some to argue that LFI (in its current narrow implementation) is part of the problem of the lack of progress on patient safety, rather than part of the solution (Cook, 2013; Kellogg et al., 2017).

Arguably, this might be a conclusion that is debatable, and there are some promising recent attempts to make LFI work better in the NHS. The establishment of an independent investigation body, the Healthcare Safety Investigation Branch (HSIB), has opened up a new form of organisational learning based on LFI for the NHS (Macrae & Vincent, 2014). HSIB receives voluntary significant incident reports from organisations, and selects specific ones for further investigation based on national priorities and their relevance to the NHS as a whole. In the investigation process, HSIB investigators within a multidisciplinary team speak to people at the organisation, but also consider similar incidents and speak to stakeholders and experts more widely with the aim of moving beyond the specifics of the incident under investigation. So far, this appears to be a very good approach for LFI at a national level, but through its set-up as a national body HSIB will most likely have limited impact on local processes for organisational learning.

2 Rationale for Learning from Everyday Work

How can adopting a RE perspective help to support organisations with their learning processes? A detailed critique of LFI was published in one of the contributions to the special issue in *Safety Science* (Sujan et al., 2017), and in this section only a couple of key arguments are summarised. Put simply, the core argument is that LFI with its focus on events that have gone wrong learns about the wrong things (or gives only a partial and skewed account) and tends to generate a limited set of interventions that often do not "stick" because they neglect the social and informal aspects of the learning process.

The LFI process kicks in when an incident happens. By definition, something has gone wrong, and the search for root causes and contributory factors begins. Maybe an elderly patient deteriorated at home and came to harm after they had been seen by an ambulance crew who had decided the patient would not need to be taken to the hospital. Simplifying for argument's sake, the LFI approach would try to understand what contributed to this adverse event, and then suggest interventions to prevent it from happening again. Maybe the clinical skills of the paramedics were not sufficient, and they require additional training. Maybe the paramedics were unsure or unaware of the applicable protocols, and so these could be updated and disseminated to all paramedics. Maybe there could be more training.

What is missing here is an appreciation of how paramedics make these kinds of difficult decisions. Frequently, there are other patients in the community who also require an ambulance, and paramedics need to make a trade-off whether to take a patient to hospital or whether to attend to the next emergency. Hospital emergency departments are busy places, and taking patients needlessly contributes to overcrowding and puts patients at risk. Again, a trade-off is necessary. Are there supporting services available in the local community? If so, it might be safer to leave the patient at home than to take them to an already busy emergency department. A RE approach aims to understand precisely such everyday trade-offs and adaptations. The purpose of learning then changes from a search of what went wrong and how it might be prevented, to what kinds of trade-offs and adaptations clinicians make and how these might be supported. The nature of interventions from a RE perspective needs to change from barriers that target specific failure sequences (e.g. protocols and training) to broader approaches that enhance the ability to anticipate, to adapt, to monitor and to learn (the resilience "cornerstones" or resilience abilities) (Hollnagel, 2010). It can be done, for example, by fostering trust and relationships as facilitators and enablers of adaptation; or by promoting psychological safety as a mechanism for bridging the gap between work-as-imagined and work-as-done.

A corollary to this shift in focus from incidents to everyday work is that organisational learning in healthcare needs to become more social and democratic. Incident reporting systems and root cause analyses are usually owned and overseen by risk management departments or patient safety officers, with little ownership by frontline healthcare workers. However, in practice, many of the actual improvements take place in less formal settings, such as lunchtime working groups or interdepartmental teams that have formed temporarily around a common improvement objective (Sujan, 2015). In other areas of the literature, the importance of such informal communities of practice has been recognised and documented (Wenger & Snyder, 2000). Staff also need to have sufficient psychological safety to speak up and create learning in dialogue through constructive criticism of ideas and views, quite unrelated to serious incidents.

Healthcare organisations have largely failed to embrace such efforts as part of their strategies for harnessing learning and improving patient safety. Organisational learning in healthcare is still limited by the dichotomy between formal risk management efforts aimed at bringing work-as-done in line with work-as-imagined, and informal frontline efforts directed at improving everyday clinical work. RE appreciates these latter efforts and aims to embed them within the organisational learning strategy.

3 A Resilience Engineering Approach

A specific example of an approach to organisational learning in healthcare based on RE thinking is the Proactive Risk Monitoring (PRIMO) approach (Sujan, 2012). The key characteristics of PRIMO are summarised in Table 1.

Table 1 Characteristics of the PRIMO approach to organisational learning

Hassle narratives	Information about work-as-done (i.e. the tensions and contradictions, which staff experience, and the trade-offs and adaptations they make) are identified empirically based on the qualitative analysis of narratives describing problems in the work environment submitted by staff.
Participation and feedback	In order to overcome the known barriers to conventional incident reporting, staff participation is encouraged through the submission of free-text narratives. Regular feedback to staff is emphasised.
Long-term and short-term improvements	In order to maintain staff participation and to combat participation fatigue, fast and visible improvements ("quick wins") to the local work environment are an important part of the PRIMO strategy that complements its longer-term aim of strengthening resilience abilities.
Staff ownership	PRIMO recognises that organisational learning is a social and participatory process. It emphasises staff ownership of improvement interventions.

4 Hassle Narratives: Capturing Work-as-Done

On the surface, PRIMO is a very simple approach based on eliciting narratives about everyday work from frontline staff. In principle, this could be done through interviews, where staff are asked to describe their everyday work, or by observing practice supported by RE tools such as the Functional Resonance Analysis Method (FRAM) (Hollnagel, 2012). These are excellent methods for research as well as for specific improvement projects, but they can also be very time-consuming, and might not be sustainable as routine practice within an ordinary healthcare setting. An alternative way of getting at information about work-as-done is by asking staff to write down narratives and stories, which can then be analysed. People will require some guidance or a "hook" to get them started. In PRIMO, this hook is the notion of hassle. Staff are asked to write about something that caused them problems during their work over the past week or that made them think or approach aspects of work differently. In some of the hospitals where this approach was adopted, it also became known as "hassle reporting". Learning from hassle is a complement to the mandatory investigation of serious adverse events. When something goes wrong, and a patient is harmed, there are issues around responsibility, accountability and blame that need to be carefully navigated. On the other hand, people experience hassles on a daily basis, and they are usually more than happy to share their hassles with people who are willing to listen. Importantly, though, analysis of hassle narratives provides useful insights into work-as-done. When people report hassles in their narratives, they frequently do not simply stop at saying things such as "*we were short of staff...*". Instead, the narratives typically continue with an account of what happened next, "*...and then I told other departments to expect delays, and I rearranged workflows...*". In this way, elicitation of hassle narratives is an excellent and very simple way of gaining deeper insights into the daily tensions and contradictions that people face (the hassles), and the trade-offs and adaptations they make. A brief example is shown below.

Example of a Hassle Narrative from a Pharmacy Setting:
"*The lead technician made me aware that the CT scanner had been down and there were 37 patients waiting for an appointment, if the scanner was fixed later today, we may see an impact. This would increase the workload on an already busy day. I told the lead clinician that I'd chase this up with [the Clinical Director] to find out if there was anything we could do to prepare for this*".

The hassle narrative provides an example of how a technician and a pharmacist create shared awareness of a problem to support anticipation of potential follow-on implications, and how these might be best dealt with. From a RE perspective, potential solutions might focus on identifying and strengthening ways of supporting the creation of shared awareness and corresponding information flows. This contrasts with more traditional interventions that might focus on increasing the reliability of the CT scanner.

5 Participation and Feedback: Overcoming the Barriers

The literature on learning from incidents describes numerous barriers to reporting, including lack of feedback, difficult incident reporting forms and systems, and lack of time (Benn et al., 2009; Lawton & Parker, 2002; Macrae, 2015). PRIMO addresses these barriers by encouraging staff participation and feedback. PRIMO aims to make it as simple as possible to contribute safety information, to give ownership over improvements to staff, and to focus on both short-term as well as long-term improvements. As opposed to structured incident report forms that force the user to adhere to a specific reporting format, hassle narratives can be submitted in people's own styles without the need for a structured form or template. They can be submitted in different formats, for example, electronically via a web platform, via email, or as a written piece of paper for those who do not have easy access to computers at work. While there is normally a designated PRIMO lead who will collect the hassle narratives and have a first look at them for themes (or to de-identify sensitive issues), this lead person is a member of staff within the specific unit or department rather than an (more or less) anonymous analyst within the risk management department. After the initial analysis, feedback is given to staff in team and departmental meetings, where issues can be discussed further.

6 Long-Term and Short-Term Improvements: Combating Participation Fatigue

A major weakness of current incident reporting systems is that they produce little actual change (Shojania, 2008). The perceived lack of learning and absence of relevance to the local work environment may have a detrimental impact on the willingness of staff to contribute to incident reporting (Firth-Cozens et al., 2004; Shojania, 2008). The result is that over time less and less useful information is generated from incident reporting systems as staff settle into a pattern of reporting only that which they know needs to be reported for bureaucratic and governance reasons. This means that the same things get reported over and over again (e.g. patient falls), but no new information becomes available, and no learning takes place (Macrae, 2015).

PRIMO addresses this participation fatigue by encouraging staff to focus not only on strategic longer-term improvements, but *also* on "quick wins", that is, fairly simple improvements that can be made within a short period of time. Quite obviously, quick wins (or low-hanging fruits) are seldomly the answer to complex problems. This has been well documented in the literature. It has been recognised that many of the numerous and well-intentioned local quality improvement projects do not result in sustainable improvements, and that they might be ineffective of addressing problems at a systems level (Dixon-Woods & Pronovost, 2016; Illingworth, 2015). Many stubborn issues require understanding of dependencies across departmental boundaries and can be addressed only through collaboration and more fundamental change at the organisational or institutional level. Why, then, the focus on the simple improvements, the quick wins? It is in the nature of longer-term improvements that they take months or longer to get off the ground. In the meantime, staff who contributed safety information typically see no changes within their local work environment. This is where the importance of these quick wins in PRIMO comes in, because they contribute to generating the momentum and the positive culture that is required for putting in place sustainable longer-term improvements. Even small local changes, such as having designated spaces for equipment that often goes missing, can make significant contributions towards maintaining staff engagement while strategic longer-term interventions are designed, implemented and evaluated.

7 Staff Ownership: Making Learning a Social Process

Ownership for organisational learning frequently is allocated to a department, for example, the risk management or clinical governance department. These departments collect, analyse and distribute safety information. Accordingly, learning from incidents has been described and criticised as focusing narrowly on dissemination of safety information without proper consideration of professional practice (Margaryan et al., 2017). Several writers have suggested that organisations need to reframe learning as a social and participative process that facilitates informal ownership of improvements, and collective sensemaking and reflection (Lukic et al., 2012; Macrae, 2015; Stanton et al., 2017).

PRIMO aims to support the social infrastructure for learning by fostering staff ownership and supporting communities of practice (Wenger & Snyder, 2000). A crucial element of PRIMO is that improvement interventions are not imposed from outside, but are generated within local meetings, and responsibility and authority for leading on specific improvements are given to staff volunteers who are closely associated with the particular process or pathway under consideration. In this way, learning can occur within the department and is directly linked to clinical practice.

Building communities of practice can be supported in various ways. Frequently, communities of practice arise spontaneously around lunchtime working groups of enthusiastic individuals working collaboratively on issues of common interest, which often cross departmental boundaries. Another approach to support communities of practice within PRIMO might be the use an electronic platform with social media functions to collect and discuss hassle narratives (see Fig. 1 for an anonymised example populated with information from a pharmacy study site).

The PRIMO approach is based on these four principles described above, but it is not enshrined further in prescriptive implementation details. This is because every department is different and will have different preferences and requirements. PRIMO can be adopted in different ways as long as the main principles are maintained.

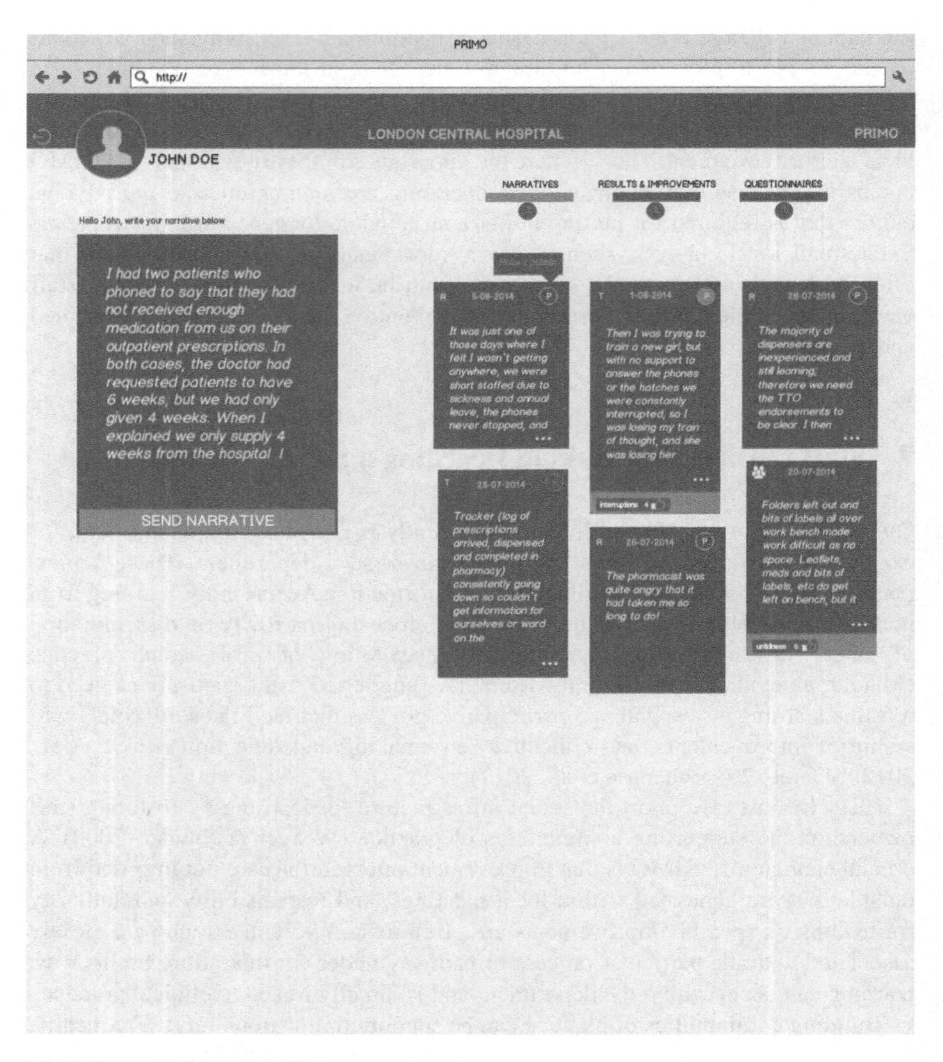

Fig. 1 Web interface for PRIMO

8 Case Studies

PRIMO Evaluation Study The PRIMO approach was developed in collaboration with one hospital as part of the Safer Clinical Systems programme (funded by the Health Foundation, a UK charity). Following the successful pilot study, the approach was then rolled out within 10 hospitals, two of which were selected for in-depth study and evaluation of the approach. The hospitals were provided with an introduction to the approach during a workshop. The two evaluation sites were visited on a monthly basis over the course of the 18-months study. During these meetings, the researcher acted as a critical friend and advisor, but project teams were free to implement and use the approach in a way that seemed most applicable and fruitful in their environment. The reason for taking this approach was twofold. On the one hand, it was recognised that learning from hassle would need to be tailored to the specific setting. On the other hand, it seemed prudent to test whether such an approach could function without close supervision by an external expert, that is, whether non-specialists could own and run the approach.

Study Sites The two evaluation study sites were English NHS hospitals. PRIMO was implemented in the Radiology department of the first hospital (site A), and in the Surgical Emergency Admissions Unit (SEAU) at the second hospital (site B). The two departments were chosen to reflect different characteristics: on the one hand a highly structured diagnostic services environment, and on the other hand a busy and dynamic ward environment that provides emergency services also during the night-time.

The radiology department at site A consists of the main X-ray department and a number of specialist modalities such as CT (computed tomography), MRI (magnetic resonance imaging) and nuclear medicine. The whole department employs approximately 90 staff. Some of these are employed part time. The roles within the department range from clerical, radiographic assistant, assistant practitioners, radiographers, specialist radiographers, advanced practitioners and consultants.

For the purpose of the study, the focus was on the main X-ray department, rather than the specialist modalities. Within the main X-ray department, there are four general rooms with a fast throughput of patients ranging from fully mobile to immobile, seriously ill patients. Referrals come from a wide range of areas, including A&E, GPs, outpatient clinics and hospital wards. There are also two specialist rooms where interventional procedures are performed. Throughout a typical working day, approximately 350 examinations are performed.

The SEAU at site B is part of the Emergency Assessment Unit (EAU), which houses also medical emergency assessment services. There are 24 beds available on EAU. EAU has a large team of medical, surgical, nursing, clerical and housekeeping staff. Referrals come from a wide range of areas, including ED, GPs, and outpatient clinics. There are between 600–800 admissions to SEAU per month. Doctors working in SEAU are not based on the ward, but are there on a rotational basis during their on-call period.

Data Collection and Analysis Data were collected from the individuals closely involved at each site, and from interviews with a wider range of staff. During the study, the implementation lead at each site kept an implementation diary. After the implementation period, in-depth interviews were conducted with members of the implementation teams (see Table 2). The implementation diaries and the interviews were analysed qualitatively through Thematic Analysis to identify what was done, any barriers and obstacles encountered, and successes achieved.

Table 2 Interview participants from the implementation teams

Study site	Role	Participant ID
A	Head of Radiology	A/IL-01
A	Radiographer (Implementation Lead)	A/IL-02
B	Surgical Trainee (Implementation Lead)	B/IL-01
B	Research Nurse	B/IL-02
B	Staff Nurse (PRIMO champion)	B/IL-03
B	Junior Doctor	B/IL-04

Table 3 Interview participants by phase and role (site A)

Pre-intervention		Post-intervention	
ID	Role	ID	Role
A01	Radiology Assistant	A10	Radiographer
A02	Radiographer	A11	Senior Radiographer
A03	Assistant Practitioner	A12	Assistant Practitioner
A04	Radiographer	A13	Radiographer
A05	Assistant Practitioner	A14	Radiographer
A06	Radiographer	A15	Radiographer
A07	Radiology Assistant	A16	Assistant Practitioner
A08	Medical Secretary	A17	Radiographer
A09	Radiographer		

Table 4 Interview participants by phase and role (site B)

Pre-intervention		Post-intervention	
ID	Role	ID	Role
B01	Ward Sister	B11	Healthcare Assistant
B02	Matron	B12	Staff Nurse
B03	Acute Care Practitioner	B13	Healthcare Assistant
B04	Clinical Educator	B14	Staff Nurse
B05	Foundation Year 1 Doctor	B15	Foundation Year 1 Doctor
B06	Foundation Year 2 Doctor	B16	Foundation Year 2 Doctor
B07	Foundation Year 2 Doctor	B17	Ward Sister
B08	Foundation Year 1 Doctor	B18	Staff Nurse
B09	Staff Nurse		
B10	Healthcare Assistant		

Semi-structured interviews with staff prior to the implementation of PRIMO and after the implementation period were undertaken to describe their safety-related attitudes and behaviours, and to determine any changes over the study period (see Table 3 and Table 4).

9 Learning: Practical and Social

Both study sites identified and implemented a range of improvement interventions. Many of these interventions addressed the reported hassles directly. For example, in response to missing and misplaced equipment in X-ray investigation rooms, an intervention based on lean thinking was developed to improve housekeeping. Similarly, on the surgical emergency admissions unit, drip stands went frequently missing as colleagues from other wards borrowed these and never returned them. A colour-coding scheme was devised to allow easy identification of drip stands that belong to the admissions unit. Arguably, such improvement interventions are not ground-breaking nor specific to RE. However, they empower people to contribute to improvements in their work environment, and they provide visible feedback that staff ideas and concerns are taken seriously.

The more profound impact of PRIMO was on the frequently neglected social and informal aspects of learning. The PRIMO approach to understanding and learning from work-as-done provided a vehicle to staff to discuss, share ideas and – importantly – engage with colleagues across departmental boundaries. For example, one of the main strategic improvement activities in the radiology site was around addressing the communication with theatres requesting radiographers to support ongoing surgery with mobile imaging equipment. This communication is time-critical and was felt to be difficult. Requests for radiographers often come in at short notice and are frequently not coordinated as they can originate from different specialities. As a result, the main radiology department might be left without appropriate cover and without appropriate supervision arrangements for junior members of staff. There might also be delays in performing the imaging in the theatre because there is only a limited number of mobile machines available, and this can cause delays in surgery and affect patient outcomes. Communication across departmental boundaries in a hospital is never an easy matter to address due to differing priorities and unclear allocation of responsibility. However, with the evidence generated from the analysis, the radiology team felt well prepared to initiate a dialogue with the theatre manager to raise awareness of this issue. Subsequently, an electronic booking diary and a standard operating procedure for booking the mobile imaging equipment were developed. This was supported by an interdepartmental working group – a community of practice – which was established specifically for this purpose.

10 Conclusion

There is broad agreement that organisational learning in healthcare is a key mechanism for improving patient safety, but at the same time frustrations with existing approaches based predominantly on learning from incidents are running high, fuelled by lack of progress and staff disengagement with learning processes. This chapter argued that current LFI processes in healthcare focus their learning on the wrong things (i.e. things that go wrong), and that they neglect the social dimensions of learning.

The chapter described an approach to learning based on RE principles that focuses on everyday work, and some of the learning that was generated as a result of running this approach in study hospitals was presented. Looking only at the practical and more tangible improvements, such as improved housekeeping, colour coding of equipment, an electronic booking diary, and a standard operating procedure, one might ask how this moves beyond existing approaches. This is a valid question, but there is a danger that one approaches learning based on RE with the same expectations and measures as one would use to assess LFI. In LFI as applied in healthcare settings, the development and implementation of such practical improvements (or safety barriers) are, in many cases, the only purpose. Within RE, the focus is on resilience abilities, and the impact on the social dimension of learning is, arguably, more important than the specific improvement interventions. Hence, we need to consider whether and how an organisation's abilities to anticipate, to adapt, to monitor and to learn have been affected. In this chapter, it was argued and attempted to demonstrate that the proposed RE approach to learning from everyday work has stimulated staff participation in the learning process, has created ownership for learning among staff, and has furthered the formation of communities of practice that are able to build relationships and dialogue to improve patient safety.

References

Argyris, C., & Schon, D. A. (1996). Organisational learning II: Theory, method and practice. Reading, MA: Addison-Wesley.

Anderson, J. E., Kodate, N., Walters, R., & Dodds, A. (2013). Can incident reporting improve safety? Healthcare practitioners' views of the effectiveness of incident reporting. *International Journal for Quality in Health Care, 25*(2), 141–150.

Benn, J., Koutantji, M., Wallace, L., Spurgeon, P., Rejman, M., Healey, A., & Vincent, C. (2009). Feedback from incident reporting: Information and action to improve patient safety. *Quality & Safety in Health Care, 18*(1), 11–21.

Braithwaite, J., Westbrook, M. T., Travaglia, J. F., & Hughes, C. (2010). Cultural and associated enablers of, and barriers to, adverse incident reporting. *Quality & Safety in Health Care, 19*(3), 229–233.

Brennan, T. A., Leape, L. L., Laird, N. M., Hebert, L., Localio, A. R., Lawthers, A. G., Newhouse, J. P., Weiler, P. C., & Hiatt, H. H. (1991). Incidence of adverse events and negligence in hospitalized patients. *New England Journal of Medicine, 324*(6), 370–376.

Carroll, J. S., & Edmondson, A. C. (2002). Leading organisational learning in health care. *Quality and Safety in Health Care, 11*(1), 51–56.

Carruthers, I., & Phillip, P. (2006). *Safety first: A report for patients, clinicians and healthcare managers*. NHS.

Cook, R. (2013). Resilience, the second story, and progress on patient safety. In E. Hollnagel, J. Braithwaite, & R. Wears (Eds.), *Resilient health care* (pp. 19–26). Ashgate.

Department of Health. (2000). *An organisation with a memory: Building a safer NHS for patients*. National Health Service.

Dixon-Woods, M., & Pronovost, P. J. (2016). Patient safety and the problem of many hands. *BMJ Quality and Safety, 25*(7). https://doi.org/10.1136/bmjqs-2016-005232.

Drupsteen, L., & Hasle, P. (2014). Why do organizations not learn from incidents? Bottlenecks, causes and conditions for a failure to effectively learn. *Accident Analysis and Prevention, 72*, 351–358.

Easterby-Smith, M., Crossan, M., & Nicolini, D. (2000). Organizational learning: Debates past, present and future. *Journal of Management Studies, 37*(6), 783–796.

Firth-Cozens, J., Redfern, N., & Moss, F. (2004). Confronting errors in patient care: The experiences of doctors and nurses. *AVMA Medical & Legal Journal, 10*(5), 184–190.

Hollnagel, E. (2010). Prologue: The scope of resilience engineering. In E. Hollnagel, J. Paries, D. D. Woods, & J. Wreathall (Eds.), *Resilience engineering in practice: A guidebook*. Ashgate.

Hollnagel, E. (2012). *FRAM, the functional resonance analysis method: Modelling complex socio-technical systems*. Ashgate.

Hollnagel, E., Wears, R.L., & Braithwaite, J. (2015). From safety-I to safety-II: A white paper.

Illingworth, J. (2015). *Continuous improvement of patient safety: The case for change in the NHS*. The Health Foundation.

Jacobsson, A., Ek, Å., & Akselsson, R. (2012). Learning from incidents–A method for assessing the effectiveness of the learning cycle. *Journal of Loss Prevention in the Process Industries, 25*(3), 561–570.

Kellogg, K. M., Hettinger, Z., Shah, M., Wears, R. L., Sellers, C. R., Squires, M., & Fairbanks, R. J. (2017). Our current approach to root cause analysis: Is it contributing to our failure to improve patient safety? *BMJ Quality & Safety, 26*(5), 381–387.

Kohn, L. T., Corrigan, J. M., & Donaldson, M. S. (2000). *To err is human: Building a safer health system*. National Academies Press.

Lawton, R., & Parker, D. (2002). Barriers to incident reporting in a healthcare system. *Quality & Safety in Health Care, 11*(1), 15–18.

Lindberg, A.-K., Hansson, S. O., & Rollenhagen, C. (2010). Learning from accidents – What more do we need to know? *Safety Science, 48*(6), 714–721.

Littlejohn, A., Margaryan, A., Vojt, G., & Lukic, D. (2017). Learning from incidents questionnaire (LFIQ): The validation of an instrument designed to measure the quality of learning from incidents in organisations. *Safety Science, 99*, 80–93.

Lukic, D., Margaryan, A., & Littlejohn, A. (2010). How organisations learn from safety incidents: A multifaceted problem. *Journal of Workplace Learning, 22*(7), 428–450.

Lukic, D., Littlejohn, A., & Margaryan, A. (2012). A framework for learning from incidents in the workplace. *Safety Science, 50*(4), 950–957.

Macrae, C. (2015). The problem with incident reporting. *BMJ Quality and Safety, 25*, 71–75.

Macrae, C., & Vincent, C. (2014). Learning from failure: The need for independent safety investigation in healthcare. *Journal of the Royal Society of Medicine, 107*(11), 439–443.

Margaryan, A., Littlejohn, A., & Stanton, N. A. (2017). Research and development agenda for learning from incidents. *Safety Science, 99*, 5–13.

National Advisory Group on the Safety of Patients in England. (2013). *A promise to learn – A commitment to act*. National Health Service.

Peerally, M. F., Carr, S., Waring, J., & Dixon-Woods, M. (2016). The problem with root cause analysis. *BMJ Quality & Safety, 26*, 417–422.

Rollenhagen, C., Alm, H., & Karlsson, K.-H. (2017). Experience feedback from in-depth event investigations: How to find and implement efficient remedial actions. *Safety Science, 99*, 71–79.

Shojania, K. G. (2008). The frustrating case of incident-reporting systems. *Quality & Safety in Health Care, 17*(6), 400–402.

Stanton, N. A., Margaryan, A., & Littlejohn, A. (2017). Editorial: Learning from incidents. *Safety Science, 99*, 1–4.

Sujan, M. A. (2012). A novel tool for organisational learning and its impact on safety culture in a hospital dispensary. *Reliability Engineering & System Safety, 101*, 21–34.

Sujan, M. (2015). An organisation without a memory: A qualitative study of hospital staff perceptions on reporting and organisational learning for patient safety. *Reliability Engineering & System Safety, 144*, 45–52.

Sujan, M. A., Huang, H., & Braithwaite, J. (2017). Learning from incidents in health care: Critique from a Safety-II perspective. *Safety Science, 99*, 115–121.

Tucker, A. L., & Edmondson, A. C. (2003). Why hospitals don't learn from failures: Organizational and psychological dynamics that inhibit system change. *California Management Review, 45*(2), 55–72.

Vincent, C., Aylin, P., Franklin, B. D., Holmes, A., Iskander, S., Jacklin, A., & Moorthy, K. (2008). Is health care getting safer? *BMJ, 337*, a2426. https://doi.org/10.1136/bmj.a2426.

Wears, R., & Sutcliffe, K. (2019). *Still not safe: Patient safety and the middle-managing of American medicine*. Oxford University Press.

Wenger, E. C., & Snyder, W. M. (2000). Communities of practice: The organizational frontier. *Harvard Business Review, 78*(1), 139–146.

Westbrook, J. I., Li, L., Lehnbom, E. C., Baysari, M. T., Braithwaite, J., Burke, R., Conn, C., & Day, R. O. (2015). What are incident reports telling us? A comparative study at two Australian hospitals of medication errors identified at audit, detected by staff and reported to an incident system. *International Journal for Quality in Health Care, 21*(1), 1–9.

Reflections on the Experience of Introducing a New Learning Tool in Hospital Settings

Sudeep Hegde and Cullen D. Jackson

Contents

Event reporting systems are widely prevalent across healthcare organizations and are used as tools to learn about a variety of negative outcomes and near misses. As such, they are artifacts of the traditional approach to safety, to learn from how things go wrong based on specific episodes or incidents. These systems typically involve self-reporting by staff involved in the incident, in compliance with policies and guidelines on reportable events. Generally, however, the effectiveness of such tools in improving safety has been limited. One of the primary reasons is that this approach to organizational learning primarily focuses on errors, near misses, and adverse events, all of which represent things that go wrong. Among the workforce, this can result in fear of blame, reprimand, and associated social and socio-legal consequences (Anderson et al., 2013; Ashcroft et al., 2006; Sujan, 2015; Waring, 2005). From a learning perspective, retrospective analysis of negative occurrences or outcomes is fraught with hindsight bias, where the adverse or potentially adverse consequence leads to a tendency of the analyst to undervalue the contextual factors that influenced or necessitated the course of actions taken prior to the event (Cook

S. Hegde (✉)
Clemson University, Clemson, South Carolina, United States

C. D. Jackson
Beth Israel Deaconess Medical Center, Harvard Medical School, Boston, MA, USA

C. P. Nemeth, E. Hollnagel (eds.), *Advancing Resilient Performance*,
https://doi.org/10.1007/978-3-030-74689-6_6

et al., 1998; Wears & Cook, 2004). As a corollary, there is an underappreciation of the contextual factors that influence adaptive responses. As a result, there is a widening gap between work as done in actual operational contexts and work as imagined by policy makers and system designers (Hollnagel, 2015, 2016).

In contrast, the resilience engineering (RE) approach is to learn from how things go well in everyday work. This is based on the premise that things go right and wrong for, essentially, the same reasons, that is, variability in performance within a variable environment. However, there is a lack of formal mechanisms and tools to operationalize such learning in organizations. Much of the current empirical literature on resilience has involved research investigators observing, interviewing, or surveying domain stakeholders. These efforts have largely been in the context of research for academic purposes. There are few, if any, examples of concerted efforts by organizations to implement frameworks for proactive learning about everyday work with the lens of resilience engineering. This chapter focuses on the development and efforts to implement a self-reporting tool for frontline caregivers at hospitals – the Resilience Engineering Tool to Improve Patient Safety (RETIPS). RETIPS is designed to enable caregivers to share narratives of adaptive performance in their everyday work. In contrast to traditional incident reporting systems, a key feature of this tool is that it aims to elicit examples of successful adaptive performance in the context of specific events as well as 'normal' routine functioning when there are no 'events'. This marks a shift toward proactively learning about normal work as it happens, including how and why performance in daily routines varies, and why it usually succeeds (when there is no event). The chapter will provide a summary of the development, initial implementation and results. The chapter will also dwell on the authors' experience in their attempt to implement it at a large multispecialty hospital.

1 Developing the Original RETIPS

The project had its origins in an interview-based knowledge elicitation technique designed to learn about how things go well in everyday clinical work. The interview was semi-structured, based on the format of the Critical Decision Method (CDM) by Klein et al. (1989). The questions were adapted to focus on situations with positive outcomes in terms of patient safety, as well as formal and informal practices, routines and adaptive measures employed in everyday work. Questions were also derived from Hollnagel's Resilience Analysis Grid (RAG) and adapted to suit the healthcare domain and clinical areas to which participants belonged. An initial study developed and conducted interviews of frontline caregivers. Qualitative analysis findings were used to develop a self-reporting form, RETIPS, which retained the essence of the interview protocol in terms of its knowledge elicitation goals. However, the structure was adapted to self-reporting, that is, a combination of free text and multiple-choice questions. Feedback was sought from experts in human factors and safety. Additionally, feedback was sought from domain stakeholders, including clinicians. The feedback at this stage is mostly related to the semantics and relevance of the content. There was not much emphasis on making the tool pragmatic, such as its length. A detailed description of the original version, including its development from the interview results, has previously been published (Hegde et al., 2015).

2 Development of RETIPS 2.0

The initial version of RETIPS was reviewed by anesthesiologists at a large multispecialty hospital as a potential tool for lesson-sharing in anesthesia. Iterative feedback was used to refine the tool, resulting in RETIPS 2.0 – a much more concise version of the original, adapted to anesthesia residents. The clinicians acknowledged the conceptual basis of the tool as relevant and were supportive of implementation on a trial basis. The feedback at this stage mostly focused on the design of the tool for practical use in clinical settings. Specific feedback included:

Conciseness: The clinicians, almost unanimously, agreed that the original version of RETIPS was too long, which would be a deterrent for potential respondents given the highly busy environment in which they work. In order to make the tool more practical, the general suggestion was to make the tool as short and concise as possible. One clinician provided a specific guideline: "it should take no longer than 10 minutes to submit a response".

Focused narrative: Several clinicians suggested customizing the tool for specific clinician groups, such as anesthesia residents, and focusing on specific safety and quality issues, such as difficult airway management. This approach would drive more focused recall and narrative. The guiding examples, cues, and response choices should be tuned accordingly. This strategy would have the added benefit of enabling analysis of patterns related to each group and issue through multiple reports. On the flipside, it could entail significant time and effort to develop separate versions of the tool for the various issues.

Clear purpose: It was important to clearly communicate to the respondent how the information provided would be used to enhance patient care. This could be done both during dissemination of the tool and in the introduction section of the tool itself.

3 Tool Description

RETIPS-AnRes consists of multiple sections, described below in sequence.

Introduction: A short paragraph is included at the beginning of the tool to define 'resilience' in a health care context and the purpose of the tool.

Case Selection: The respondent is asked to think of examples from their own work practice that relate to resilience in terms of preventing patient harm. This question field guides the respondent to think of either of two types of examples: a specific instance or a generic routine or process.. The examples were chosen so as to be relevant to anesthesia workflows and, therefore, more relatable to the respondent.

Detailed Narrative: This field is provided for the respondent to describe in detail the 'resilience' example they considered in the previous section. The following thematic cues are provided to guide the respondents with their descriptions: Key Challenges and Concerns; Adaptive Responses; Anticipation; Preventive Measures; Monitoring Behaviors (checks, reviews etc.); Resource Availability; Policies and Standard Practices; Communication; Cooperation; Patient/Family Involvement. There is no suggested word limit to the narrative description.

The remaining sections of RETIPS consist of checkbox-type responses intended as probes on various aspects (e.g., success factors, challenges, resources) of the reported case that are relevant from a resilience perspective. Each response field includes a text box to allow respondents to elaborate or describe other factors not listed that may have been involved in their example.

What Went Right: This field is designed to highlight the factors that contributed to success or the factors that were favorable to the goals inherent in the resilience example being related. Response choices include experience and knowledge of co-workers; culture and attitudes; standard practice/policy; shared understanding; cooperation between co-workers; and leadership.

Challenges and Concerns: This section more specifically probes the issues that challenged or threatened patient safety, or impeded successful intervention. The six response choices are: patient condition or behavior; communication issues; complexity of the situation; uncertainty or ambiguity; limited resources; and policy issues.

Resources: This section asks the respondent to check off those resources that were useful in the situation(s) they previously described: adequate time; technology/equipment; co-workers/consults; information; and procedural guidelines.

Area of Practice: The respondents are asked to indicate the specific clinical area of practice related to their example, such as surgical, preoperative, and postoperative anesthesia.

4 Implementation of RETIPS 2.0

Pilot Implementation: After multiple revisions, RETIPS-AnRes was administered twice in a two-year period to consecutive batches of anesthesia residents in their internship year (first postgraduate year, or PGY-1). The residents served as a representative group of the potential user-population, that is, frontline caregivers. The tool was implemented as part of a week-long course on Quality Improvement (QI) for each cohort of anesthesia interns. RETIPS-AnRes was welcomed by the residency program directors as it could serve as a tool that would allow for self-reflection on challenges and triumphs during the residents' workday. Such reflection is key for a number of professional milestones outlined by the Accreditation Council for Graduate Medical Education (ACGME): self-directed lifelong learning (ACGME Anesthesia PBLI Milestone 3: Self-directed Learning) as well as System-based Practice Milestone 3 (Patient Safety & QI), and Professionalism Milestones 3 (Commitment to Institution Department and Colleagues) and 4 (Receiving & Giving Feedback). These milestones are a way in which the ACGME determines who can practice medicine at a defined level of proficiency through a competency-based model (Holmboe et al., 2016). Therefore, RETIPS-AnRes was incorporated into the QI course curriculum as a self-reporting exercise for the interns. The tool was made available in an electronic format through REDCap, a Health Insurance Portability and Accountability Act (HIPAA)-compliant web-based application used to build and manage online surveys (Harris et al., 2009). In order to encourage use of the tool, the department's residency program director offered extra credits for submission of responses. Responses to the tool were anonymized. Reports were reviewed by authors, SH and CDJ, for overall response patterns and alignment of responses with the purpose of the tool.

Additionally, the tool was also made available to all anesthesia residents at the hospital. Dissemination strategies included emails to the residents introducing the tool, its purpose and potential impact. The emails were endorsed by the chief residents as well as the residency program directors. Unfortunately, however, no response was received.

In year one, nine reports were received from the six residents, and in year two, four reports were received. As the reports were anonymous, it was not possible to tell whether all six residents responded in year one or whether any participant submitted more than one report in year two. Participants briefly described lessons learned, and indicated the success factors, challenges, and resources pertinent to their examples. Four of the responses were categorized by the participants as 'specific', four as 'generic', and one as 'both'. However, upon further examination of the responses given, we found that three of those categorized as 'generic' were, in fact, specific instances of routine workflows. Some examples of the responses collected are given below:

Exemplar Response 1: "During this robotic assisted case, despite a low probability of requiring blood products during this OR case, the anesthesia resident made sure to have blood readily available in room in the event that an adverse event occurred during which the daVinci robot could not be moved out of the surgical field quickly. The rationale was that in case a vital artery (specifically, the pulmonary artery in this lung wedge resection) is injured during the robot assisted surgery, blood could be given rapidly using units in the room as a temporalizing measure rather than having to wait for blood products to be transferred from blood bank while the robot could be moved away from the surgical field."

- *Category:* Specific.
- *Success Factors:* Experience and knowledge of co-workers; standard practice/policy; cooperation between co-workers; shared understanding; culture and attitudes.

- *Comment:* "Communicating with the circulating nurse, OR nurse and blood bank allowed for blood units to be available for this case, which followed from the close conversation the resident had with the anesthesia attending and with the surgical team. This allowed us to plan for and foresee potential catastrophic outcomes prior to start of the case."
- *Challenges:* Communication issues; complexity of situation. Comment: "Getting blood products into the OR seems to be a critical but oftentimes time-delayed process. Having blood on hand was an important portion of the anesthesia plan as detailed above, however despite the best efforts of the OR team blood took over 40min to make it into the room."
- *Resources:* Adequate time; procedural guidelines.

Exemplar Response 2: "During my overnight calls (especially when I am cross-covering patients from other services), I like to touch base with each one of the nurses about our patients before doing my own rounds. In this brief meeting, I like to address their concerns for the night. I feel that gathering this information before seeing the patients helps me to have more effective and productive rounds. In this interaction, I also inform the nurse about my plans during the night and I make myself available for anything during the night. I feel that this practice improves our communication and their trust level in me as the intern on service."

- *Category:* Generic.
- *Success Factors:* Experience and knowledge of co-workers; cooperation between co-workers; shared understanding; leadership.
- *Challenges:* Communication issues; complexity of situation.
- *Resources:* Adequate time; co-workers/consults; information.

Exemplar Response 3: "On positioning of patient in a prone position, all available members in the OR assist in flipping the patient. The anesthesia resident is the one in charge of communicating and directing members of the team. I observed my senior anesthesia resident clearly giving instructions on how to flip the patient and explicitly stated the order to put the monitors back on the patient in order of importance. The patient was not flipped until all members of the team were ready. Care was taken with all IV lines and with the ET tube. The patient had become hypotensive on induction, so care was taken to closely monitor blood pressure after patient positioning. Care was also taken to relieve areas of pressure points. Arms were placed in the neutral position."

- *Category:* Specific.
- *Success Factors*: Experience and knowledge of co-workers; cooperation between co-workers; shared understanding; culture and attitudes; leadership.
- *Challenges:* Limited resources.
- *Comment:* "One of the arm boards of the bed didn't seem to be working and thus the patient was not positioned appropriately. This was recognized and instructions were given to nursing to help to retrieve another arm board. The surgical team assisted with patient positioning and ultimately the patient's arms were placed in a satisfactory position."

- *Resources:* Adequate time; technology/equipment; co-workers/consults.

Exemplar Response 4: "If there are no surgeries booked at a certain time cutoff the acute care surgery operating room becomes available for other services so that surgeries such as transplants can be performed. This assures prompt attention for these patients minimizing ischemic times, etc."

- *Category:* Both.
- *Success Factors*: Experience and knowledge of co-workers; standard practice/policy; cooperation between co-workers; shared understanding; culture and attitudes; leadership.
- *Challenges:* Uncertainty or ambiguity in the situation; limited resources; policy issues.
- *Resources:* Information.

5 Reflections on the Pilot Implementation Experience

Based on the responses submitted to RETIPS 2.0, it seems that the expectations of the tool were generally understood by the respondents. This was generally reflected in the fact that the responses ranged from specific incidents or episodes to generic routines free of any episodic context. However, all responses illustrated elements of variability in the environment and performance. Participants were able to explicitly articulate elements of variability in their everyday work in a variety of contexts. Therefore, the tool seems to be effective in terms of its knowledge elicitation objective. That said, given the limited number of responses, their usefulness in informing the identification of any systemic patterns of adaptation cannot be verified yet. Aggregating a sufficient number of examples of variability in a specific area could allow for such patterns to be identified.

However, the larger challenge that became evident from the exercise related to the uptake of the tool in operational settings. The healthy response rate – 13 reports from 12 participants – from the cohort of residents who were part of the quality improvement training can be attributed to the fact that RETIPS was included as part of their curriculum with extra credit for submitting reports. There was a clear purpose and incentive for fulfilling course goals, within a time-bound context. In sharp contrast, there were no responses received from the larger resident population despite a formal endorsement by faculty who were residency directors, and follow up circulars encouraging residents to submit responses. This is not surprising given the many competing priorities for their time, including clinical duties and fulfilling educational requirements. Another reason may be that there was no clear perceived direct and immediate benefit to the respondents. For instance, it would not have been apparent to the resident whether submitting the reports would help be followed by improvement actions or policy changes in the near term, or not.

6 Development and Towards Implementation of RETIPS-Airway Management

In line with the feedback provided by experts and clinical leaders earlier, a third version was developed which focused on a specific issue in anesthesia – airway management or difficult intubation. The previous version of the tool was adapted by replacing examples with those relevant to airway management and modifying questions and response choices to be more specific to the clinical issue. These modifications were made by involving anesthesiologists and nurse anesthetists over multiple iterations to ensure relevance and coherence of the content.

The original strategy of the authors was to incorporate the RETIPS-Airway management form into the hospital's existing event-reporting portal. The portal consists of separate forms for various safety issues and clinical areas, such as patient falls and blood transfusion. The idea of RETIPS as a way to learn proactively about how things go well in everyday work was supported by the Vice President for Health Care Quality at the hospital. However, an important question that emerged in terms of implementation was what resources would be required to administer, maintain and process reports once they would be generated, and how would such a project be funded. The authors assured the leadership that no additional resources would be necessary if the tool were to be incorporated within the existing portal infrastructure, and that reports would initially be used primarily for analysis and research purposes. Additionally, a significant logistical issue was encountered in that the portal had a set template for forms, which could not accommodate the structure and full content of RETIPS. Therefore, as a compromise, the authors had to relinquish the idea of including RETIPS-Airway Management as a separate form within the portal. Instead, we included two key questions from RETIPS to the portal's existing airway-management form meant for adverse event or near-miss reporting. The questions focused on expanding beyond the specific event to describe how the process usually goes well: (1) "Context beyond this incident, what usually goes well?" and (2) "Please use the text box to describe workflows, decisions, and factors that enable effective airway management and risk prevention under the *usual* circumstances." Technical constraints inherent to the portal's design meant that responses to these questions could not be made mandatory for form submission. Only five responses to the questions were received out of 45 reports over the span of 9 months, July 2018 to April 2019. None of the responses provided information about care under usual circumstances or what usually goes well, but instead amplified information about the event being reported.

7 Reflections on the Overall Experience of Applying RETIPS in a Hospital: What Went Well and What We Learned

As mentioned earlier, the design of the tool itself was effective in terms of knowledge elicitation about everyday variability in performance, and the key factors that contribute to the challenges and successes of performance.

Buy-in from leadership, engagement with key stakeholders: A strong and sustained intent from the hospital's clinical and safety leadership is important for driving a new initiative, more so if it involves a shift in thinking. Therefore, it is necessary to communicate, not just the idea of RETIPS, but the approach to learning that it represents, to various levels of leadership in the hospital. In our effort to implement RETIPS hospital-wide, we met with the Vice President for Health Care Quality at the hospital and described the idea of a tool designed to learn proactively about 'normal' work, rather than just adverse events; they were immediately agreeable to exploring how the tool could be implemented. This meeting led to further discussions with other key stakeholders in the organization, such as the Clinical Manager for Perioperative Education, who reports to the Associate Chief Nurse for Perioperative Services. The endorsement of high-level administrators and managers enabled us to reach out to other key stakeholders, such as technical and administrative staff, whose support was imperative to operationalizing RETIPS. We do not have an answer yet to the best possible way to get staff to engage on a wide scale in the learning process. However, communicating the ideas demonstrated in the tool to stakeholders at multiple levels could be a useful way to start. We have found that the idea of RE and learning from how things go well is a simple yet compelling concept. There was no disagreement encountered at the conceptual level. However, this does precipitate questions on the 'how-tos' regarding operationalizing the concept.

Using Organizational 'hooks' to operationalize the tool: Our approach was not to propose a replacement to existing reporting and learning systems at the hospital, but to influence existing workflows and learning pathways in the organization. To this end, we identified existing processes in the hospital and department as organizational 'hooks', to 'latch' RETIPS onto. Again, communicating to key stakeholders, the purpose of the tool and what we were looking for in terms of implementation, was crucial in identifying such hooks. For instance, in order to pilot the tool, we spoke to the residency supervisors in the Anesthesia department. In turn, they suggested the quality improvement course as a forum to introduce RETIPS to the residents through their curriculum. At the hospital level, the existing event reporting portal was identified as a platform to introduce questions from RETIPS. Similarly, appropriate organizational hooks could be identified for integration of the tool or its parts in concert with clinical and administrative groups at departmental and hospital levels.

Incentive to respondents: In our experience, offering extra course credit for submitting reports seemed to help generate a relatively high response rate (13 reports from a total of 12 residents) within the course participation. In general, however, obtaining a healthy response rate for meaningful analysis remains a significant challenge. When implementing the tool more widely, beyond the ACGME course work, it may be useful to offer a suitable incentive to targeted respondents, especially initially, in order to drive responses. Promotion by senior leadership or supervisors is essential.

Confidentiality: RETIPS reporting should be confidential in nature so as to protect the identity of respondents. Confidentiality is important to help respondents feel secure about discussing any potentially sensitive aspects of their work, including risks and informal workarounds. Confidentiality, rather than anonymity, would also enable analysts and investigators to follow up with the respondent to gather additional details related to their report, and engage them in any subsequent improvement efforts. This might have contributed to responses from residents. We did receive more than the minimum number.

Follow-up, analysis, and feedback: A key factor in sustaining engagement of the workforce with reporting is communicating the outcome or impact of the reports to the respondents. When the analysis and its ensuing decisions are made visible to the organization, staff are motivated to continue to report as they see their responses as being impactful on their environment. This feedback loop should, therefore, be an essential part of the larger organizational learning framework of which RETIPS itself would be a part. Furthermore, the exemplar responses submitted through RETIPS do not provide all details to understand a work practice or the environment, but can be used as triggers for further investigation by safety administrators, senior management, hospital analysts, and even clinicians. In addition, these examples could be used to seed a survey asking for an expansion of the list of examples to a wider audience. This has the virtue of not requiring someone to complete it immediately after cases and may result in richer input since respondents should not be as time constrained. While we have not done this in our implementation so far, the visibility of information from individual reports back to the professional community has been acknowledged as one of the main factors underlying the sustained success of the Aviation Safety Reporting System (ASRS) (Cook et al., 1998).

In addition to the above reflections, we offer the following takeaway from our experience: In order to successfully implement RETIPS, it was important for us to identify networks of people and agencies at various organizational levels, who can influence and/or may be influenced by the change. Identifying the network also entails understanding the relationships in these networks. In this regard, relevant questions to consider include: who works for whom (hierarchical and lateral dynamics), what are their usual responsibilities and scope, how are resources shared, what would it require people to do in order for the proposed changes to be implemented, and are there perceived benefits? Engaging with various levels of the workforce helped us not only understand the system and its networks better, but also to communicate our ideas more effectively. Among other benefits, this process of continuous engagement and dialogue enabled the identification of the aforementioned organizational hooks, which are key to implementation. This approach follows from the core RE principle of learning about normal work to implementing changes that blend with the flow of normal work.

8 RETIPS in the Context of a Larger Organizational Learning Framework

RETIPS is designed as an artifact of the 'new thinking' towards proactive learning in resilient systems. The organizational hooks mentioned earlier are but a few ways in which RETIPS was introduced into the hospital's learning processes. Going forward, RETIPS could be part of a larger organizational learning framework, complementing other information-sharing pathways by emphasizing the focus on learning proactively about how things go well in everyday work. Other organizational hooks that could be opportunistically leveraged include:

1. Departmental Morbidity and Mortality (M&M) Conferences: These could be used to describe and discuss examples of everyday performance variability, including those identified through RETIPS.
2. Combined Safety Grand Rounds: This is a forum where the entire Perioperative community (periop services, surgery, anesthesia, ortho surgery, OB/GYN) come together for a lecture or activity on patient safety. Some of these could be devoted to clinical resilience, and may be a way to incentivize staff to provide 'stories of resilience' that would be called out in this forum.
3. Simulation: Debriefs during simulation exercises could be focused more on what went well rather than just areas for improvement in the technical and nontechnical work so that individuals and teams can discuss aspects of their actions and workflows which contribute to safe patient care.
4. Lecture: Numerous opportunities are available, especially at teaching hospitals, with residents, staff, and students to introduce them to concepts of resilience and how they relate to everyday clinical work, quality of care, and patient safety.

As mentioned in the 'Reflections' sections earlier, the leadership would have to be involved in facilitating reporting and conversations with RETIPS through the existing learning pathways. Appropriate incentives can be identified based on the type of staff (e.g., residents, attendings, nurses, technicians), area or clinical specialty, organizational level, and other factors. The course credit for residents used in our pilot implementation is an example. Other incentives can include formal recognitions and awards. Data from various areas of the hospital can be analyzed under a common analytic framework with a 'resilience lens'. Such analysis would involve identifying patterns, including adaptations, resource usage, and communication. These patterns further inform the recognition of resource needs, process redesign requirements, policy changes, etc. for various concept-driven goals, such as reducing brittleness and improving adaptive capacity.

9 Conclusion

RETIPS, a previously developed lesson-sharing tool based on Safety-II and Resilience Engineering principles was revised and reconfigured, based on feedback from clinicians, in terms of relevance and practicality within the hospital setting. The revised version of the tool, RETIPS-AnRes, was disseminated on a pilot basis to anesthesia interns as part of the curriculum of a one-week course on quality improvement. The implementation validated the design in that the responses were aligned with the purpose of the tool, which was to learn about how things go well in everyday clinical work. Further, the tool was adapted for a specific clinical issue – difficult intubation and airway management. Reflecting on the overall experience of implementing RETIPS, we summarize key takeaways for operationalizing the tool in hospital settings. This work demonstrates the potential for RETIPS as a means for proactive organizational learning in healthcare, widening the focus beyond adverse events and near misses. The potential for wider and longer-term implementation of RETIPS within a larger organizational framework for learning about resilience in frontline medical work is also discussed.

References

Anderson, J. E., Kodate, N., Walters, R., & Dodds, A. (2013). Can incident reporting improve safety? Healthcare practitioners' views of the effectiveness of incident reporting. *International Journal for Quality in Health Care, 25*, 141–150. https://doi.org/10.1093/intqhc/mzs081.

Ashcroft, D. M., Morecroft, C., Parker, D., & Noyce, P. R. (2006). Likelihood of reporting adverse events in community pharmacy: An experimental study. *Quality and Safety in Health Care, 15*, 48–52. https://doi.org/10.1136/qshc.2005.014639.

Cook, R. I., Woods, D. D., & Miller, C. (1998). *A tale of two stories: Contrasting views of patient safety*. Chicago, IL: National Patient Safety Foundation.

Harris, P. A., Taylor, R., Thielke, R., Payne, J., Gonzalez, N., & Conde, J. G. (2009). Research electronic data capture (REDCap)—a metadata-driven methodology and workflow process for providing translational research informatics support. *Journal of biomedical informatics, 42*(2), 377–381.

Hegde, S., Hettinger, A. Z., Fairbanks, R. J., Wreathall, J., Wears, R. L., & Bisantz, A. M. (2015). Knowledge elicitation for resilience engineering in health care. In *Proceedings of the Human Factors and Ergonomics Society Annual Meeting* (pp. 175–179). https://doi.org/10.1177/1541931215591036.

Hegde, S., Hettinger, A. Z., Fairbanks, R. J., Wreathall, J., Krevat, S. A., Jackson, C. D., & Bisantz, A. M. (2020). Qualitative findings from a pilot stage implementation of a novel organizational learning tool toward operationalizing the Safety-II paradigm in health care. *Applied Ergonomics*, 82. https://doi.org/10.1016/j.apergo.2019.102913.

Hollnagel, E. (2015). Why is work-as-imagined different from work-as-done? In R. L. Wears, E. Hollnagel, & J. Braithwaite (Eds.), *The resilience of everyday clinical work. Resilient Health Care* (Vol. 2, pp. 249–264). Ashgate.

Hollnagel, E. (2016). Prologue: Why do our expectations of how work should be done never correspond exactly to how work is done. In J. Braithwaite & R. L. Wears (Eds.), *Reconciling work-as-imagined and work-as-done. Resilient health care III* (pp. 7–16). CRC Press.

Holmboe, E. S., Edgar, L., & Hamstra, S. (2016). The milestones guidebook. Chicago, IL: Accreditation Council for Graduate Medical Education.

Klein, G. A., Calderwood, R., & Macgregor, D. (1989). Critical decision method for eliciting knowledge. *IEEE Transactions on systems, man, and cybernetics, 19*(3), 462–472.

Sujan, M. A. (2015). An organisation without a memory: A qualitative study of hospital staff perceptions on reporting and organisational learning for patient safety. *Reliability Engineering and System Safety, 144*, 45–52. https://doi.org/10.1016/J.RESS.2015.07.011.

Waring, J. J. (2005). Beyond blame: Cultural barriers to medical incident reporting. *Social Science & Medicine, 60*, 1927–1935. https://doi.org/10.1016/j.socscimed.2004.08.055.

Wears, R. L., & Cook, R. I. (2004). The illusion of explanation. *Academic Emergency Medicine, 11*, 1064–1065. https://doi.org/10.1197/j.aem.2004.07.001.

Resilient Performance in Aviation

Meredith Carroll and Shem Malmquist

Contents

Resilience is defined as "the intrinsic ability of a system to adjust its functioning prior to, during, or following changes and disturbances, so that it can sustain required operations under both expected and unexpected conditions" (Woods & Hollnagel, 2006, p. xxxvi). If there was ever an industry that has demonstrated this ability, it is the aviation industry. As has been evident in recent events ranging from the 737 MAX debacle to the COVID-19 pandemic, the aviation industry is incredibly sensitive to a broad array of disturbances ranging from those that are aviation-specific such as accidents that wreak havoc on consumer perceptions, to economic and environmental circumstances that can bring air travel to a screeching halt. The industry has continually demonstrated the ability to adjust and sustain operations after unexpected events, such as September 11 (Blunk et al., 2006) and the major recession in 2008 (Franke & John, 2011). Although less obvious, there is also evidence that aviation has proactively adjusted its functioning prior to disturbances to maintain safe and effective operations. The aviation industry has improved both reliability and safety in the midst of increasing complexity of the aircraft, economic challenges, and aviation systems that are dependent on a range of different organizations to succeed (Høyland & Aase, 2008). It has been proposed that resilience is a characteristic of system performance, not the system itself (Hollnagel, 2011), and therefore it is fitting to examine the aspects of aviation that enable it to demonstrate resilient performance. This chapter presents a discussion of resilient performance in

M. Carroll (✉) · S. Malmquist
Florida Institute of Technology, Melbourne, FL, USA
e-mail: mcarroll@fit.edu

C. P. Nemeth, E. Hollnagel (eds.), *Advancing Resilient Performance*,
https://doi.org/10.1007/978-3-030-74689-6_7

aviation, including what resilient performance looks like in aviation, how it is currently achieved, and methods to further advance resilient performance in the future.

1 What is Resilient Performance in Aviation?

At the heart of aviation's ability to demonstrate resilient performance is the performers on which the industry relies most heavily: aircraft pilots. Pilots are considered the fundamental safety component when aircraft systems do not operate as expected, and the assumption is that pilots will be able to anticipate and recover after encountering a problem for which the aircraft systems were not designed (NTSB, 2019). Prior to the advent of computers and automation, the type of unexpected events that pilots encountered were often the result of gaps in our knowledge of the physical world, such as the unknown effects of supersonic flow over a wing (NASA, 2008) or some aspect of weather such as a microburst event (Caracena et al., 1986), or limitations of our ability to perceive and comprehend relevant cues from the environment (e.g., the horizon). The increase in automation has fundamentally changed the piloting task. Once a correspondence task, in which pilots experienced cues and determined how they correspond to previous experiences in order to gain situational awareness, the piloting task is now a coherence task, in which pilots consume information provided by automated systems to gain this understanding (Mosier & Fischer, 2010). This has resulted in a change in the type of unexpected events for which pilots must anticipate and be prepared to respond. For example, the Boeing 737 MAX accidents in which flawed data from an angle of attack sensor led the flight control computers to determine that the aircraft was at a dangerously high angle, resulted in the aircraft computers responding in a way that was unexpected and for which the pilots had not been trained (FAA, 2019). There are, however, unexpected events such as these, from which pilots managed to recover. For example, in 2012, an EVA Air Airbus A330 experienced an un-commanded pitch down due to iced-over angle of attack vanes. The pilot recovered "with only seconds to spare" by turning off all three air data reference systems, a procedure that was not trained and had never been performed before by anyone (Lambregts, 2013, p. 1368). What is it that differentiated the outcome of these events? It was the pilots' ability to leverage their experience and problem-solve, in order to respond to, and recover from, an unexpected event.

Hale and Heijer (2006) define resilience as the ability not only to recover from an adverse event, but also the ability to anticipate and adjust in order to prevent adverse events. With respect to aviation, this hinges on pilots having the knowledge, skills, abilities, and resources to anticipate unexpected events such as those discussed previously, so that they can make an effective decision regarding how to prevent and/or respond. These are often events for which they have not received training or procedures. It also requires that the systems, and team members available to support pilots in the larger aviation system, are able to anticipate and respond to the changing needs of the pilots. Carroll et al. (2012) propose a model for resilient decision making that is both in-line with Hale and Heijer's (2006) definition and can be adapted to shed light on resilient performance in aviation. Carroll et al. propose that resilience, what we refer to herein as resilient performance, consists of two components: (a) an initial phase characterized by the need to adapt performance to prevent or minimize the impact of the adverse event, and (b) a second phase characterized by recovery from the adverse event. The first phase, in which there is the need to adapt, determines the level of performance decrements from which the performer must recover in the second phase (See Fig. 1). When individuals are highly adaptable and effectively adjust performance, performance decrements are minimized and the performer has less ground to make up in the recovery phase, if any at all. When performers successfully anticipate an unexpected event and effectively adjust performance, they are able to maintain performance and sustain required operations. Carroll et al.'s model is based on the pathway model of human resilience and findings that human response to a significantly stressful event can range from succumbing to the stressor and performance falling apart, to surviving with degraded performance, to fully recovering, or even thriving (Carver, 1998; Bananno & Mancini, 2012). This model served as a basis for development of a resilience classification algorithm to quantify individual resilience to acute stress, and researchers were able to identify which trajectory a performer was most likely to follow based on baseline physiological and self-report measures (Winslow et al., 2015).

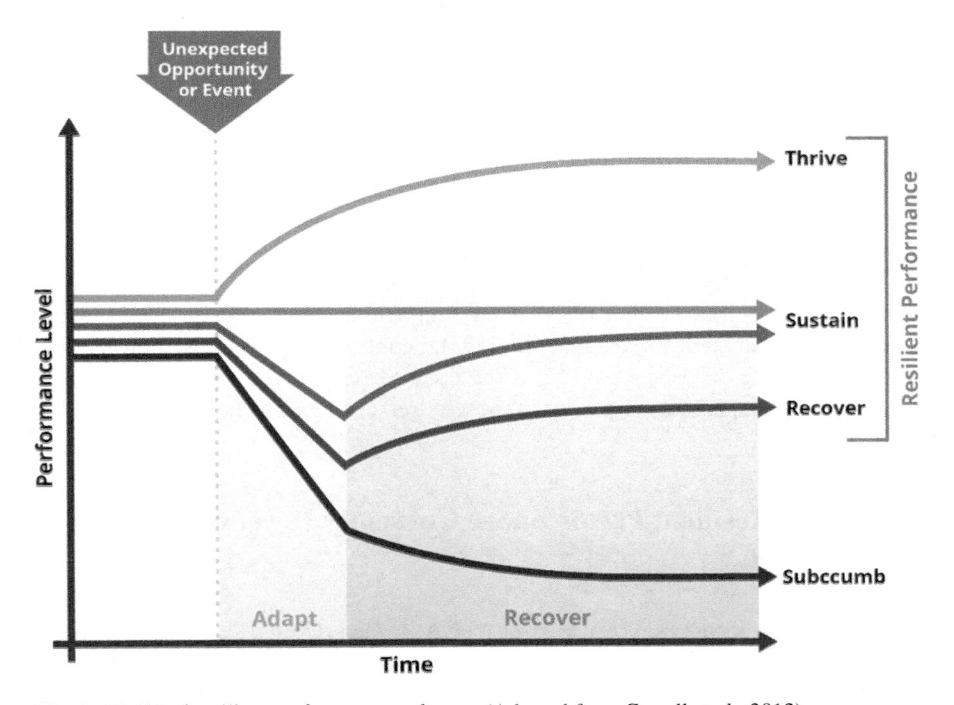

Fig. 1 Model of resilient performance pathways (Adapted from Carroll et al., 2012)

Although different in many ways, there is similarity in an individual performer's response to an acute stressor and an unexpected event. In both cases, the more adaptable an individual is, and the better able the individual is to anticipate, cope, and recover, the more likely their performance trajectory is to realign with baseline levels of performance and state. This is the goal of a system in Woods and Hollnagel's (2006) definition: "to sustain required operations under both expected and unexpected conditions" (p. xxxvi). This model also incorporates a trajectory that aligns with a later definition of resilient performance that incorporates adjusting, not only to changes and disturbances, but also to opportunities (Hollnagel, 2015). In this trajectory, a performer or system capitalizes on an unexpected opportunity to boost or advance performance levels.

What influences which trajectory will eventually result? Hollnagel (2011) proposes that there are four key processes that enable resilient performance and they include the ability to: (a) monitor information relevant to system performance and the surrounding environment, (b) anticipate potential disruptions, demands and opportunities, (c) respond to disturbances and opportunities by adjusting performance, and (d) learn how future performance should be adjusted based on observations and experiences. These processes, and the degree to which a performer or system is set up to accomplish these processes, will differentiate whether a performer can maintain system functioning, recover or thrive, or succumb to the disturbance resulting in performance suffering. An example of this model playing out in an aviation context can be seen in the previous example of the EVA Airbus A330 incident. The A330 has fly-by-wire flight controls that feature a system utilizing the angle of attack sensors that will automatically prevent an aerodynamic stall by lowering the aircraft nose. In this incident, icing led to the angle of attack sensors incorrectly indicating a stall and the automatic system continuing to lower the nose without improvement. The pilots quickly adapted to the situation, by shutting down all three air data reference computers, which forced the computers to change to a degraded mode that deactivated the stall prevention system (Kaminski-Morrow, 2019). This was not a process for which the pilots had a procedure or training. The pilots, based on their understanding of the aircraft systems, and the monitoring of the system performance and relevant environmental cues (e.g., out the window view) successfully determined that the stall indication was false. This allowed them to respond to the disturbance by adjusting their performance. The industry, having learned from this, added the procedure to required training for the A330.

2 How is Resilient Performance Currently Developed in Aviation?

Although it has been a trial by fire, over the last several decades, the aviation industry has built up a hefty safety management system designed to facilitate resilient performance. The key to this has been the utilization of data-driven approaches that allow the industry to *monitor* unexpected events and disturbances, and trends in performance that allow the industry to anticipate future disturbances. The industry has been able to learn from this data in order to shape pilot training and procedures in very effective ways, in order to prepare pilots to know how to respond to and anticipate unexpected events and maintain effective performance.

The aviation industry collects vast amounts of data related to flight performance and safety, from a range of sources that are both reactive in nature (e.g., accident analyses, safety reporting systems), and proactive in nature (e.g., system data and performance monitoring programs; Congress, 1988). The first large-scale attempt to capture this data was with the Aviation Safety Reporting System (ASRS), developed by NASA under the leadership of Charles Billings (Billings et al., 1976). ASRS allows pilots to voluntarily report incidents or safety concerns, while being protected against any punitive action associated with the event. The ASRS reports are available via a searchable database, which researchers and safety professionals can utilize to identify situations in which latent hazards exist and/or near-accidents have occurred. Similar internal company-specific flight safety reporting programs, such as the Aviation Safety Action Program (ASAP), are also widely utilized within aviation to capture additional safety data that can be accessed and utilized in analysis (Cusick et al., 2017). The Federal Aviation Administration's (FAA) Aircraft and Flight Operations Quality Assurance (FOQA) program captures quantitative data directly from the aircraft sensors such as airspeeds, altitudes, descent rates, accelerations, headings, and flight control positions (FAA, 2007). FOQA allows airlines to identify trends and safety events not otherwise reported. Additionally, the Line Oriented Safety Audit (LOSA) program involves the use of trained observers who ride in the cockpit and record their observations during flights to identify trends in how pilots apply procedures on the flight deck and areas in which deviations occur (Cusick et al., 2017).

This multifaceted approach to data monitoring provides the aviation industry with an enormous amount of rich data about the types of events, expected and unexpected, for which performers need to be prepared to anticipate and equipped to respond. Information from these databases is extracted and analyzed in a range of different ways to identify risks or trends such as pilots extending flaps at too high of a speed, risky, or ineffective approaches, or unsafe aircraft-loading policies (Cusick et al., 2017; FAA, 2018). Based on these trends, aircraft manufacturers, regulatory agencies, and airlines can work to put mitigations in place, such as the redesign of systems, policies, procedures, and training, to prepare aviation performers to adjust functioning in anticipation of, or in response to, future disturbances (International Civil Aviation Organization, 2013). For example, not long after implementing an ASAP program, one major U.S. airline discovered a large number of reported altitude deviations due to communication procedures, and after training the pilots on a new communication procedure, the problem disappeared (National Business Aviation Association, 2019). In another example, a problem in the airport approach procedure for the Orlando International Airport was forcing pilots to be too high on approach, resulting in unstable approaches and long landings; the crossing altitude was changed for part of the arrival alleviating the problem (FAA, 2007). In another example, ASAP reports revealed a risk of a possible runway overrun during takeoff at San Francisco International Airport, and a Safety Alert for Operators was issued to mitigate the risk, resulting in airlines modifying their training and procedures (National Business Aviation Association, 2019).

Monitoring relevant information sources and learning both from unexpected events and general performance trends are powerful tools that enable resilient performance. Within the aviation industry, resilient performance has been achieved by developing programs and processes that facilitate: (a) the monitoring of relevant events and data, (b) the analysis of this data in order to learn from the events and data, (c) review of procedures and policies related to events and performance trends in order to anticipate future issues, and (d) the adapting of training and safety programs and policies to prepare performers to respond to future events and states (International Air Transport Association, 2013).

3 Advancing Resilient Performance in Aviation

While this data-driven approach has proven successful at facilitating resilient performance in aviation, there is always an opportunity for advancement. The key to achieving this is moving beyond the data. There is a need to not only update training and procedures to prepare pilots for the unexpected events that can now be anticipated due to past experiences. We must develop the ability of our pilots to have the foresight and flexibility to adjust performance in the face of a truly unanticipated event, for which there is no training, procedures, or any expectation; and to be able to anticipate the potential for such an event and adapt to prevent occurrence. We need to integrate, into the aviation industry's safety management approach, training and procedures that specifically enable an aviator's ability to anticipate, adapt, and recover from a range of completely unexpected events. Leveraging Woods and Hollnagel's (2006) definition of resilience as a goal, and Carroll et al.'s (2012) model of resilience as a framework, there is an opportunity to reinforce the abilities necessary to achieve resilient performance in pilots, and the aviation industry at large. Here, we provide a few ways in which such an approach could be implemented.

With respect to the first phase in Carroll et al.'s model in which performers must adapt, there is an opportunity to bolster pilots' ability to utilize inductive reasoning by incorporating training that focuses more heavily on problem-solving within ill-defined events. Currently, pilot training is primarily focused on procedures – normal procedures, emergencies procedures, checklists, and protocols (Rapoport & Malmquist, 2019). Procedures are key for ensuring safe and consistent operations across a broad range of situations. Thus, pilot training is heavily focused on honing pilot performance within the rule-based and skill-based modes of Rasmussen's model of operator performance (Rasmussen, 1983). To achieve this, pilots repeatedly practice the application of procedures, or rules, and the performance of psychomotor skills necessary to develop incredibly reliable performance under normal, and abnormal but predicable (e.g., emergency), performance conditions. As a result, pilots develop keen skills in detecting, troubleshooting and responding to predictable situations. However, resilient performance maintains safe operations not only under nominal and standard, off-nominal conditions, but also when something totally unpredictable occurs. This situation places the performer in a knowledge-based mode of performance which requires a high degree of problem-solving and inductive-reasoning skills. Current pilot training does not focus on this ill-defined and unbounded performance domain. While some pilots have a natural aptitude and seek to develop these skills on their own accord, there is currently not a standard training regime aimed at ensuring all pilots are proficient in this performance mode (Rapoport & Malmquist, 2019). A key part of this is ensuring pilots have accurate mental models of their systems and how situations unfold (Rasmussen, 1983). With the complex automation in modern aircraft, pilots often do not possess a comprehensive understanding of how these systems work, and therefore have limited ability to anticipate unexpected events and disturbances. For example, a lack of understanding about the various flight control system modes, and how these influence the relationship between a pilot input and a control surface or throttle movement, were causal in multiple recent aircraft accidents, including Air France 447, American Airlines 587, and Asiana Airlines 214 (Rapoport & Malmquist, 2019). In order for pilots to know what system and environmental information to monitor, and how to anticipate and respond to unexpected system disturbances, it is necessary for pilots to have a thorough understanding of how their systems work. Further, pilots must also be given the opportunity to practice monitoring, anticipating, responding to, and learning from completely unexpected events.

One way to achieve this is to incorporate training scenarios that center around low-probability system failures/events, for which there is not a procedure or checklist. Currently, training focuses on failures with the highest probability of occurrence, and emergency/abnormal events such as engine failures that have the highest levels of risk associated. Pilots learn the application of checklists and procedures in these situations and practice them in the simulator to hone associated skills. Such skills are often trained to the point of automaticity, and pilots become quite good at operating in this skill-based mode. If a broad range of low probability failures could be integrated into the training regime, pilots would be provided an opportunity to practice the four abilities that Hollnagel (2011) proposes are necessary for resilient performance. Specifically, pilots could be given the opportunity to (a) monitor relevant system data and environmental cues, in order to detect unexpected anomalies in this data, (b) anticipate what this could mean for operational functioning, (c) respond to unexpected events by utilizing inductive reasoning and problem-solving skills to determine how to adjust performance to cope with the failure, and (d) learn from these events via a facilitated debrief. These training scenarios could be administered in a high-fidelity simulator to support the entire problem-solving process, including detection of relevant multimodal cues, recognition of what they mean for system performance, and the inductive reasoning which must ensue to make sense of the current situation and path forward. However, the later stages of recognition and problem-solving could be easily practiced in less-expensive, lower fidelity simulations such as Tactical Decision Games (TDGs).

TDGs are a technique utilized by the military to rapidly and inexpensively expose trainees to numerous situations (Crichton et al., 2000). They provide the opportunity to make decisions and receive feedback on the course of action chosen, in order to build up the experience base from which they can draw during actual performance. Such a technique could be leveraged to increase a pilot's experience base, and to provide opportunities to practice inductive reasoning and problem-solving. For example, a TDG scenario for pilots might include a verbal description of an en route situation, followed by a verbal description and/or graphical representation via a handout or PowerPoint slide representing a sudden change in the aircraft state as indicated by the flight deck displays, instruments, and other multimodal cues. The pilot(s) could be given time to ponder the situation, determine what failure(s) have most likely occurred, and select the best course of action. An instructor could then facilitate a structured discussion regarding the inductive reasoning process that the pilot(s) utilized and provide guidance on honing this process. Such an approach could be utilized in a group setting in which pilots are given the opportunity to problem-solve individually, and then participate in a group discussion. This technique could continue with additional scenarios, in which pilots could be given decreasing amounts of time to respond, with the goal of increasing the efficiency of the inductive reasoning processes they perform. If pilots can be trained to more effectively adapt to an unexpected situation or failure, then performance decrements might be minimized, thereby minimizing the recovery required to regain baseline performance.

With respect to bolstering the recovery phase, there is an opportunity to prepare pilots to respond more effectively under stress. Pilots must be trained to manage their internal systems, in addition to the external system of which they are a part. This includes monitoring, recognizing, responding to, and recovering from the negative psychological, physiological, and performance impacts resulting from unanticipated events. Unanticipated events often cause a stress response within a performer and although individual differences exist, research has shown clear patterns regarding the impact that stress has on a performer's psychological, psychological, and decision-making response (McNeil & Morgan, 2010). This is seen in landmark accidents such as United 173 in which the pilots let the aircraft run out of fuel while troubleshooting a landing gear problem (NTSB, 1978). Such attentional narrowing is known to result as a response to stress (Staal, 2004), and pilots must be trained to anticipate and compensate for these decrements. The military has recognized the need to prepare individuals to perform under stress, and has responded by developing targeted stress training. For example, the Infantry Immersion Trainer provides warfighters an opportunity to make decisions under highly realistic stressors prior to deployment (Muller et al., 2008). By providing performers the opportunity to experience a realistic stress response, performers can learn to recognize what happens to their physiology and decision processes and develop mechanisms for coping with these challenges. The aviation industry has not integrated this approach into their training regime. Although pilots practice performing under time constraints and the moderate stress of a challenging check ride, they do not currently have the opportunity to respond to a critical and completely novel situation on the flight deck while their heart is pounding and their chest is tight. Many pilots will be required to perform under these conditions at some point in their career. As such, it is critical that a pilot is able to effectively maintain operational functioning under these conditions.

One way to achieve this is by incorporating training approaches that induce significant levels of stress and require pilots to problem-solve and make decisions under these circumstances. There are studies which have shown the efficacy of stress training techniques in pilot training (McClernon et al., 2011); however, it is challenging to induce high levels of stress in a training setting. The use of stress induction techniques such as social evaluative stressors has been shown to result in significant stress response during simulation training exercises, with participants from relevant samples such as the military (Carroll et al., 2014). Specifically, Carroll et al. (2014) designed a military analogue to the Trier Social Stress Test, a highly validated stress induction technique that incorporates elements of anticipation, public speaking, and mental arithmetic (Kirschbaum et al., 1993). These three elements were operationalized based on current military training practices, making it feasible to integrate into current military training approaches. Such an approach could be implemented within pilot training, utilizing some portion of their current simulation training curriculum. For example, in the anticipation phase, pilots could be given a very limited amount of time to prepare for a flight with much more complex requirements than they are accustomed. The pilots could then be asked to brief their plan to instructor pilot(s) who will be assessing the plan, and who maintain flat affect, eye contact, and put their plan under fire. Pilots could then execute the simulation scenarios for which they have planned, encountering unpredictable and novel events for which they have no procedure or training. This could be coupled with training that provides pilots knowledge of stress impacts and coping strategies that can be utilized to mitigate the effects of stress. Such an approach could help train pilots to both recognize their stress response and learn to cope with this response in order to maintain performance. This could result in pilots being better able to recover, psychologically, physically, and with respect to performance, from system disturbances that are experienced as a result of a completely unexpected event or failure.

4 Conclusion

This chapter illustrates the great strides that the aviation industry has taken to develop resilient performers. By utilizing data-driven approaches to learn from past events, the aviation industry has continually learned from past events and adjusted training and procedures to ensure that pilots can adapt to, and recover from, a range of unexpected events and disturbances. The aviation industry has the opportunity to further advance the resilient performance of pilots by incorporating training approaches aimed to bolster a pilot's ability to problem-solve in the face of completely unexpected events, and to cope with the impact that a truly stressful situation has on their performance and state. Such approaches provide the opportunity to further advance resilient performance in aviation, leading to increased safety in the skies.

References

Billings, C. E., Lauber, J. K., Funkhouser, H., Lyman, E.G., & Huff, E. M. (1976). *NASA Aviation Safety Reporting System quarterly report.* 76, April 15, 1976 - July 14, 1976. (Report No. NASA TM X-3445). U.S. Government Printing Office.

Blunk, S. S., Clark, D. E., & McGibany, J. M. (2006). Evaluating the long-run impacts of the 9/11 terrorist attacks on U.S. domestic airline travel. *Applied Economics, 38*(4), 363–370.

Bonanno, G. A., & Mancini, A. D. (2012). Beyond resilience and PTSD: Mapping the heterogeneity of responses to potential trauma. *Psychological Trauma: Theory, Research, Practice, and Policy, 4*(1), 74.

Caracena, F., Ortiz, R., & Augustine, J. A. (1986). *The crash of Delta Flight 191 at Dallas-Fort Worth International Airport on 2 August 1985: Multiscale analysis of weather conditions.* https://repository.library.noaa.gov/view/noaa/9503/noaa_9503_DS1.pdf

Carroll, M., Hale, K., Stanney, K., Woodman, M., DeVore, L., Squire, P., & Sciarini, L. (2012). Framework for training adaptable and stress-resilient decision making. *Proceedings of the Interservice/Industry Training, Simulation, and Education Conference (I/ITSEC) Annual Meeting.* Orlando, FL.

Carroll, M., Winslow, B., Padron, C., Surpris, G., Murphy, J., Wong, J. & Squire, P. (2014). Inducing stress in warfighters during simulation-based training. *Proceedings of the Interservice/Industry Training, Simulation, and Education Conference (I/ITSEC) Annual Meeting.* Orlando, FL.

Carver, C. S. (1998). Resilience and thriving: Issues, models, and linkages. *Journal of Social Issues, 54*(2), 245–266.

Crichton, M. T., Flin, R., & Rattray, W. A. (2000). Training decision makers–tactical decision games. *Journal of Contingencies and Crisis Management, 8*(4), 208–217.

Cusick, S. K., Cortes, A. I., & Rodrigues, C. C. (2017). *Commercial aviation safety.* McGraw Hill Professional.

Federal Aviation Administration. (2007). Volume 11 Flight Standards Programs, Chapter 2-Voluntary Safety Programs, Section 2-Flight Operational Quality Assurance (FOQA). http://fsims.faa.gov/wdocs/8900.1/v11%20afs%20programs/chapter%2002/11_002_002_chg_0a.htm

Federal Aviation Administration. (2018). *Out front on airline safety: Two decades of continuous evolution.* [Fact sheet] https://www.faa.gov/news/fact_sheets/news_story.cfm?newsId=22975

Federal Aviation Administration. (2019). *Boeing 737 MAX flight control system joint authorities technical review (JATR): Observations, findings, and recommendations.* https://www. .gov/news/media/attachments/Final_JATR_Submittal_to_FAA_Oct_2019.pdf

Franke, M., & John, F. (2011). What comes next after recession?–Airline industry scenarios and potential end games. *Journal of Air Transport Management, 17*(1), 19–26.

Hale, A., & Heijer, T. (2006). Defining resilience. *Resilience Engineering.* CRC Press, 35–40.

Hollnagel, E. (2011). RAG-The Resilience Analysis Grid. *Resilience engineering in practice: A guidebook.* Ashgate.

Hollnagel, E. (2015). Introduction to the Resilience Analysis Grid. https://www.ida.liu.se/~729A71/Literature/Resilience_M/Hollnagel_2015.pdf

Høyland, S., & Aase, K. (2008). Does change challenge safety? Complexity in the civil aviation transport system. Proceedings of the *ESREL 2008 & 17th SRA Europe Annual Conference,* 22-25.

International Air Transport Association. (2013). *Evidence-based training implementation guide.* IATA.

International Civil Aviation Association. (2013). *Manual of evidence-based training.* ICAO.

Kaminski-Morrow, D. (2019). How Airbus fought its own pitch battle against rogue air data: European experience reflects key difference over 737 MAX's design and service length. *Flight International.* https://www.flightglobal.com/analysis/analysis-how-airbus-fought-its-own-pitch-battle/132358.article

Kirschbaum, C., Pirke, K. M., & Hellhammer, D. H. (1993). The 'Trier Social Stress Test'--A tool for investigating psychobiological stress responses in a laboratory setting. *Neuropsychobiology, 28*(1-2), 76–81.

Lambregts, A.A. (2013). Flight envelope protection for automatic and augmented manual control. *Proceedings of the EuroGNC*, 1364-1383.

McClernon, C. K., McCauley, M. E., O'Connor, P. E., & Warm, J. S. (2011). Stress training improves performance during a stressful flight. *Human Factors, 53*(3), 207–218.

McNeil, J. A., & Morgan, C. A. (2010). Cognition and decision making in extreme environments. In J. Moore & C. H. Kennedy (Eds.), *Military Neuropsychology* (pp. 361–382). Springer Publishing.

Mosier, K. L., & Fischer, U. M. (2010). Judgment and decision making by individuals and teams: Issues, models, and applications. *Reviews of Human Factors and Ergonomics, 6*(1), 198–256.

Muller, P., Schmorrow, D., & Buscemi, T. (2008). The Infantry Immersion Trainer: Today's holodeck. *Marine Corps Gazette, 23.*

National Aeronautics and Space Administration. (2008). *Columbia crew survival investigation report.* https://www.nasa.gov/langley/100/breaking-the-sound-barrier-fast-as-you-can

National Business Aviation Association. (2019). Voluntary reporting hazards often leads to corrective actions that make a difference in day-to-day operations. (2019). *Business Aviation Insider.* https://nbaa.org/news/business-aviation-insider/making-real-life-safety-improvements/

National Transportation Safety Bureau. (1978). *Aircraft accident report.* https://www.ntsb.gov/investigations/AccidentReports/Reports/AAR7907.pdf Rapoport,

National Transportation Safety Bureau. (2019). *Safety recommendation report.* https://www.ntsb.gov/investigations/AccidentReports/Reports/ASR1901.pdf

Rapoport, R., & Malmquist, S. (2019). *Angle of attack: Air France 447 and the future of aviation safety.* Lexographic Press.

Rasmussen, J. (1983). Skills, rules, and knowledge; signals, signs, and symbols, and other distinctions in human performance models. *IEEE Transactions on Systems, Man, And Cybernetics, 3*, 257–266.

Staal, M.A. (2004). *Stress, cognition, and human performance: A literature review and conceptual framework.* (NASA Tech. Memorandum 212824). NASA Ames Research Center.

United States Congress. (1988). *Safe skies for tomorrow: Aviation safety in a competitive environment,* (OTA-SET-381). Office of Technology Assessment, Government Printing Office, 183.

Winslow, B. D., Carroll, M. B., Martin, J. W., Surpris, G., & Chadderdon, G. L. (2015). Identification of resilient individuals and those at risk for performance deficits under stress. *Frontiers in Neuroscience, 9*(328), 1–10.

Woods, D. D., & Hollnagel, E. (2006). Prologue: resilience engineering concepts. *Resilience Engineering: Concepts and precepts, 1*, 1–6.

Assessing the Impacts of Ship Automation Using the Functional Resonance Analysis Method

Pedro Ferreira and Gesa Praetorius

Contents

The maritime industry is experiencing a steady evolution towards a concept of fully automated ship operation. The implementation and use of automated systems have been debated for many decades, and yet substantial issues remain regarding its achievements in terms of improved safety and efficiency (Wiener & Curry, 1980). The assessment of potential impacts (i.e. risk assessment) emerging from the introduction of automation remains a key challenge. The integration and streamlining of operations significantly increase complexity, and the transformations that are introduced tend to produce unforeseen side effects, often with serious safety consequences (Dekker et al., 2011).

The developments towards autonomous shipping have heavily focused on the ship side and concept developments for shore centres (e.g. Rolls-Royce Shore Control Centre), but less on how shore-based vessel operations may potentially be integrated into the current maritime transport system. It will likely require transformations, which are to have legal, economic, and organizational impacts across the industry greatly extending beyond the availability of technology. Suitably addressing these challenges requires a predictive and integrated investigation of these

P. Ferreira (✉)
CENTEC – University of Lisbon, Lisbon, Portugal

G. Praetorius
Linnaeus University, University of South-Eastern Norway, Notodden, Norway

C. P. Nemeth, E. Hollnagel (eds.), *Advancing Resilient Performance*,
https://doi.org/10.1007/978-3-030-74689-6_8

potential transformations. Particular attention should be devoted to how increased operational interdependency may generate new complexity-related aspects and how this in turn will affect the system's ability to resilient operations.

This chapter serves as a basic discussion for how the Functional Resonance Analysis Method (FRAM) can be used to explore and design functional set-ups for potential complex maritime operational scenarios. It focuses not only on the traffic management system, but also considers shore-based control centres (SCC) and other services that can be foreseen as requirements for the operation of autonomous vessels. We have identified three safety critical scenarios and described them based on different focus group activities carried out with subject matter experts. We further use the FRAM to highlight where potential future critical coordination aspects may emerge amongst different functional requirements and discuss how these may impact on the system's ability to resilient operations. The discussion builds around how the pursuit of a FRAM-based analysis of future operational concepts may contribute to enhanced resilience in increasingly dynamic and variable maritime operational conditions.

1 Maritime Traffic Management and Autonomous Vessels

Currently, navigation in and out of ports is organized as a distributed control system (Praetorius et al. 2015, Van Westrenen & Praetorius 2014). Vessels navigate according to their individual voyage plan. They may be assisted by a pilot, who is a navigational expert with specific local knowledge to increase safety of navigation, and represent the coastal state. Furthermore, coastal states often install so-called Vessel Traffic Services (VTS) in port approaches. In shore-based VTS centres, VTS Operators (VTSOs) monitor the traffic, assist in navigational matters, and provide information to all commercial vessels in a designated area, normally port areas or areas that pose navigational difficulties. It is important to note that while VTS is a support to maritime traffic, the decisions and responsibility for a vessel's safe conduct remain with the Master on board (IALA, 2016). VTSOs are thus not able to steer or direct the traffic from shore.

In the current maritime traffic system, communication from and to the VTS serves an important function as it is a source of information about the overall state of traffic and potential dangers within the VTS area. The information is public and broadcasted on dedicated radio channels that navigational crews can listen to. This in turn enables the vessels to adjust and adapt to changes as they occur. However, this is anticipated to be affected of changes to the organizational frame of traffic management once shore centres for autonomous vessel operations are introduced. Shore-control centres (SCC) are likely to represent an additional centralised control layer, which may significantly impact on the coordination resources that VTS currently ensures via its information services.

While the degree of automation in the operation of merchant vessels has steadily increased throughout the past 30 years, the maritime domain is now on the verge of a new technical revolution towards autonomous vessels, sometimes labelled shipping 4.0. (Lambou & Masaharu 2017). Recently, the International Maritime Organization has defined four levels of autonomy for Maritime Autonomous Surface Ships (MASS) (IMO, 2019):

- Ship with automated processes and decision support: Seafarers are on board to operate and control shipboard systems and functions. Some operations may be automated.
- Remotely controlled ship with seafarers on board: The ship is controlled and operated from another location, but seafarers are on board.
- Remotely controlled ship without seafarers on board: The ship is controlled and operated from another location. There are no seafarers on board.
- Fully autonomous ship: The operating system of the ship is able to make decisions and determine actions by itself.

However, despite the efforts to define a common terminology, several researchers argue that the usage of "autonomy" in the industry is somewhat misleading. While automation and autonomy are closely related concepts, the first will not automatically lead to the latter as assumed by many maritime stakeholders. As discussed by among other such as Relling et al. (2018) and Hult, Praetorius & Sandbrerg (2019), the current concepts of autonomous vessels represent a system with supervisory control from shore rather than being an autonomous actor in the traffic system.

As the degree of automation within operations increases, new perspectives are needed to explore the complexity of everyday work. Within the maritime domain, resilience engineering and its concepts have received an increasing amount of attention. Previous research has successfully applied the resilience abilities to understand everyday adaption and flexibility, as well as several researchers have modelled shore services and ship-to-shore operations to understand how safety is promoted by the various services and actors within the maritime transport system (Praetorius & Hollnagel, 2014; DeVries, 2017). The analysis and discussion in this chapter build upon earlier work and use the Functional Resonance Analysis Method to explore potential system transformations that may emerge beyond the change of its individual components (i.e. ship-board and shore-side automation, communication technologies, among others).

2 Defining Scenarios for Future Operations

Three focus group interviews were conducted to design and explore future traffic scenarios to capture, on the one hand, relevant aspects of current everyday operations (Work as Done – WAD), but also identify the transformations that are most likely to be introduced in the future, as the industry progresses towards increased automation.

The first focus group was used to develop three scenarios. Seven experts from the northern European maritime cluster representing different stakeholders in the maritime domain and academia participated in the group interview. The scenarios were developed as open-ended as possible to trigger the participants to freely discuss potential shore-based services, activities of those services, system requirements, and competence needs.

1. Reduced crew scenario

After an 8 h shift, the navigation of the vessel is handed over to a shore- based centre. The Master remains on board and can quickly be called to the bridge in case of any complications.

2. Convoy

A convoy of (unmanned) vessels is led and steered by a manned support vessel through ship–ship communication. The support vessel offers an opportunity to intervene and quickly react in case of any unanticipated events on any of the convoy's vessels. The incentive for this traffic solution was the low cost for manning. The convoy is in coastal traffic and several of the vessels are going to leave it in the approach to the next port.

3. Going to port

A vessel approaching one of Europe's major ports. During the voyage across the Atlantic, the vessel has been unmanned and steered from a shore centre. Now she is going to port to take on new cargo.

After the scenarios had been defined, two new group interviews were conducted to discuss the actual changes to and impacts of future operations. The participants were presented with the short description of the scenarios and ask to elaborate on two questions; what shore functions/activities are needed for the scenario to be realised, and who does what in terms of the identified functions and activities. The discussions were captured in terms of notes on a whiteboard to facilitate the discussions among the participants. Follow-up questions were used to explore particularly critical interactions between shore-based services and the autonomous traffic.

The experts' discussions first focused on the overall services, or service functions that would be required for operations in general. Approaching a port, even if the vessel is steered from shore during sea voyage, will require a navigator or pilot who can take the vessel to the berth. Additionally, a crew is needed for mooring and cargo handling operations. Upon arrival at the berth, linesmen and other port services, such as port authority and customs will remain a part of the infrastructure to ensure the safe and secure handling of the cargo.

It is also likely that a traffic information services, such as VTS today, will remain as shore-based function. VTS is the only service directed towards overseeing the overall traffic flow, thus efficiency and safety within a port approach. There is still a large uncertainty surrounding the coordination and communication functions between shore and ship. The participants in the two follow-up focus groups also highlighted that an unmanned sea voyage and port approach is likely to require two different SCCs, one that is focused on open sea and one centre that is area-specific, and which will take over once the vessel is approaching. Some of the local control functions could potentially be taken over by the VTS, if traffic control services become centralized in an analogy to Air Traffic Control and the VTS would gain an increased mandate beyond what is provided today. However, the participants emphasized that this would require the shipping companies to transfer some of their autonomy to services representing the coastal state, which may not be desired.

For the navigation support, a local Shore Control Centre (SCC) may be established by the shipowner in the port. The SCC will, however, not be manned by pilots or VTS operators, nor will it be the responsibility of the maritime administration to implement such a centre. Given the interest of shipowners, the centre will be manned through the shipping companies to secure the business and trade secrets.

Pilots have an important role for the safety in port approach. In today's setting, they provide three important functions to a vessel: the local language, expert ship handling in a specific area assisting manoeuvring, and they are representative of the coastal state, that is, an important safety measure to ensure safe operations in the approach. While a vessel may be unmanned and operated from shore, it is important to consider how to ensure that there is a last safeguard before a vessel can enter the port. In the current traffic system, the pilot represents this function as he or she is able to see whether a ship and crew are in the condition to enter port. Further, having to board a pilot to the vessel, especially if she is unmanned, represents additional safety risks. If there is no crew on board, the boarding procedures will have to be determined. Further, the legal implications and split of command between navigator and pilot need to be clarified. Currently, the pilot is assisting the Master, but does not have the legal responsibility and accountability for the vessel's safe voyage.

3 Modelling Future Maritime Operations

To explore future maritime operation, the functional resonance analysis method (FRAM) (Hollnagel, 2012) was used. FRAM is a method to analyse and model complex sociotechnical systems, in which functions are distributed over human operators, organizations and technology. It provides the means to model future operation concepts with a focus on overall system aspects, despite the substantial uncertainty that persists relating to the design and operation of individual system elements.

The method focuses on the concept of performance variability and ways in which systems manage and monitor potential and actual variability. FRAM is based on the principle of equivalence of successes and failures, principle of approximate adjustments, principle of emergence, and the principle of functional resonance (Hollnagel, 2012).

The modelling focused on the approach to port as one of the most safety critical scenarios of maritime operations. The analysis set-up was followed on previous exploratory work carried out in relation to automation in the context of air traffic management (Ferreira & Cañas, 2019), in which FRAM was used to investigate how foreseeable steps towards automation would impact on overall air traffic control operations. For the purpose of this exploratory work, basic assumptions were derived from the scenario description:

- VTS services are expected to be maintained under formats similar to current ones, as multiple types of "conventional" merchant vessels are expected to remain in operation within a foreseeable future.
- Interactions between the autonomous vessel approaching port, and, therefore, coming into the VTS area, and the VTS itself, will be carried out via the SCC. Since no crew is expected to be aboard, all legal and operational responsibility will necessarily be with the Officer of the Watch (OOW) in the shore-based centre.
- A pilot and minimum crew requirements are considered needed to navigate in and out of ports. Safe navigation in close interaction with a wide variety of vessels was not considered realistic under autonomous, nor remote control modes. In the approach to ports (at least major and busiest ports), a pilot and crew will, therefore, have to be dispatched and board autonomous vessels.

The maritime operations taken into account range from the approach of an autonomous vessels to a VTS area, to the manual takeover of that vessel by crew and pilot to be dispatched aboard. The model obtained is shown in Fig. 1. Different shades of grey are used to highlight three operational areas:

- VTS functions are represented in dark grey/black.
- SCC functions are represented in medium dark grey. They essentially focus on the gathering of information relating to autonomous vessels and about general traffic from VTS, and the communication, both to VTS and to autonomous vessels.
- Functions in light grey represent the operations carried out on board the vessel, once pilot and crew have boarded. These mainly relate to the necessary checks of vessel operation and systems, and all the requirements for vessel control handover.

The function represented with thinner lines (confirmation of pilot and crew) relates to pilot and crew arrangements, prior to the vessel boarding operations. Based on the model developed, two fundamental aspects of coordination are further explored:

- Interactions between SCC and VTS
- Interactions with autonomous vessel during pilot and crew boarding

These aspects of coordination were investigated through the insight on functional variability that FRAM enables. The FRAM Model Visualiser (FMV – www.functionalresonance.com) provides useful insight to investigate the "resonance effect" based on the description of the potential variability in the output of functions, with regard to its time (too early, on time, too late, or not at all) and its precision (precise, acceptable, or imprecise).

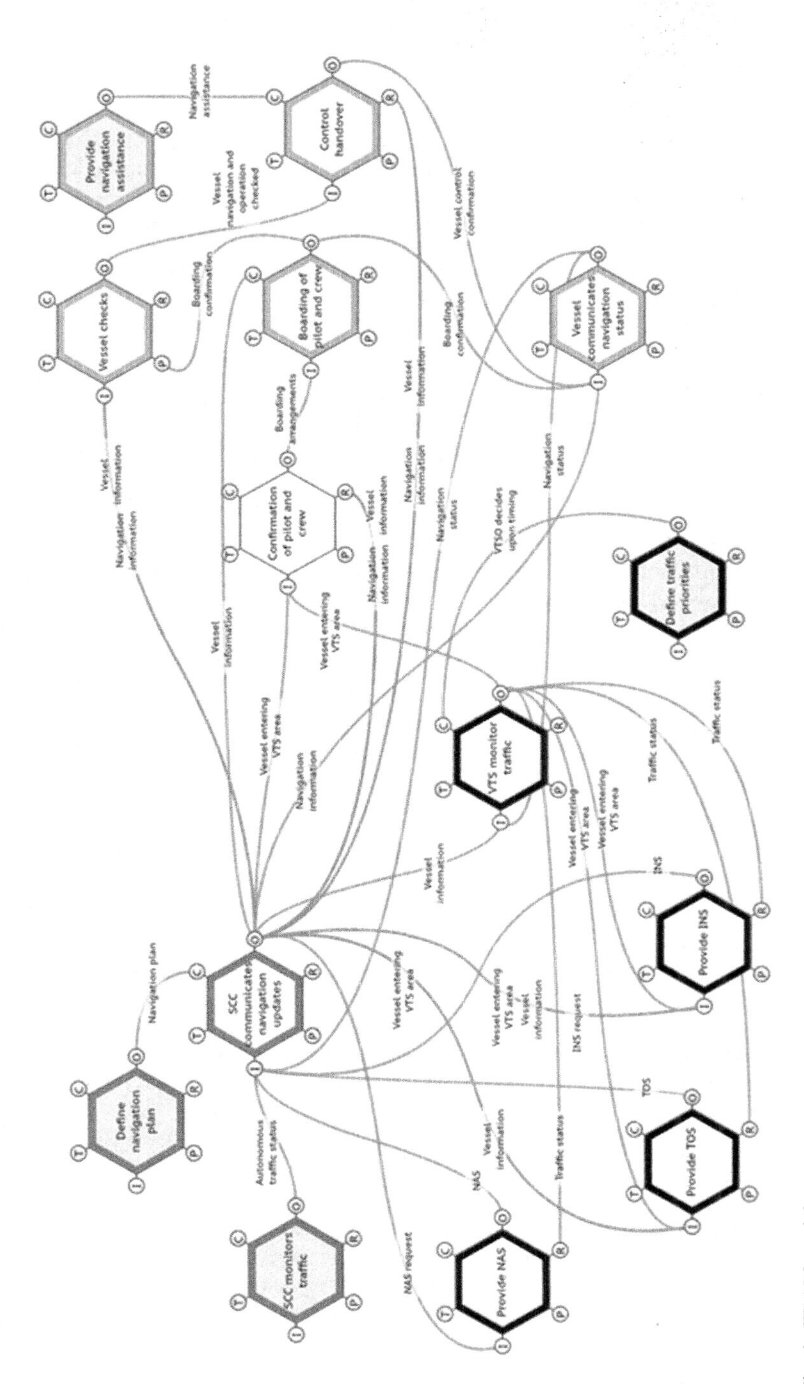

Fig. 1 FRAM model

4 Interactions Between SCC and VTS

The SCC communication is the most coupled function (SCC communicates navigation updates). This is not surprising when, to a great extent, in the scenario in question, the SCC is conveying to the autonomous vessel, information sourced through VTS services, not only to navigate the autonomous vessel (under remote control conditions), but also to assist crew as they board the vessel and make the arrangements for a control handover. As the diversity and number of vessels navigating within port areas can be expected to increase, the ability of SCC to generate a suitable overview of navigation conditions becomes increasingly limited. Hence, SCC would still rely on VTS services to develop overall traffic conditions and accordingly navigate autonomous vessels under their control. SCC is also likely to feed to VTS information and navigation data relating to the vessels under its control. This would generate the feedback loops between VTS, SCC, and autonomous vessels that are illustrated in Fig. 2.

The feedback loops in Fig. 2 also indicate that, to some extent, the VTS would need to rely on the SCC to provide the service with updated information on the state and status of autonomous vessels within their area. However, similar to current maritime regulations and procedures, the SCC is unlikely to be under obligation to provide information to the VTS, which means that these loops may not necessarily be suitably balanced. For instance, if the workload of the SCC operator becomes critical due to some particular traffic conditions, or when operating under some degraded mode, the service may withhold information from the VTS. Naturally, the VTS's ability to provide information is mainly grounded on the broader monitoring of port traffic (function "VTS monitor traffic"), but communication with traffic is an important part of their ability to generate an overview of navigation conditions and anticipate potential risks and opportunities in the traffic organization. This is where the additional centralised control layer that SCC creates, may become critical, as there is no foreseeable framework to ensure the coordination between VTS and SCC information needs.

5 Interactions Between the Autonomous Vessel and Boarding a Crew

The exchange of information between the SCC and the VTS will have a critical impact on how the boarding and handover processes will be carried out. This will most likely require a certain amount of systems check, in addition to the planning and decision-making related to navigation requirements, for which input from the VTS will be fundamental. The formal handover of vessel control between onboard crew and SCC is unlikely to be carried out before the crew aboard has completed all necessary checks, such as testing equipment and vessel response, accepting or adjusting the voyage plan made by the SCC, and has taken certain position required

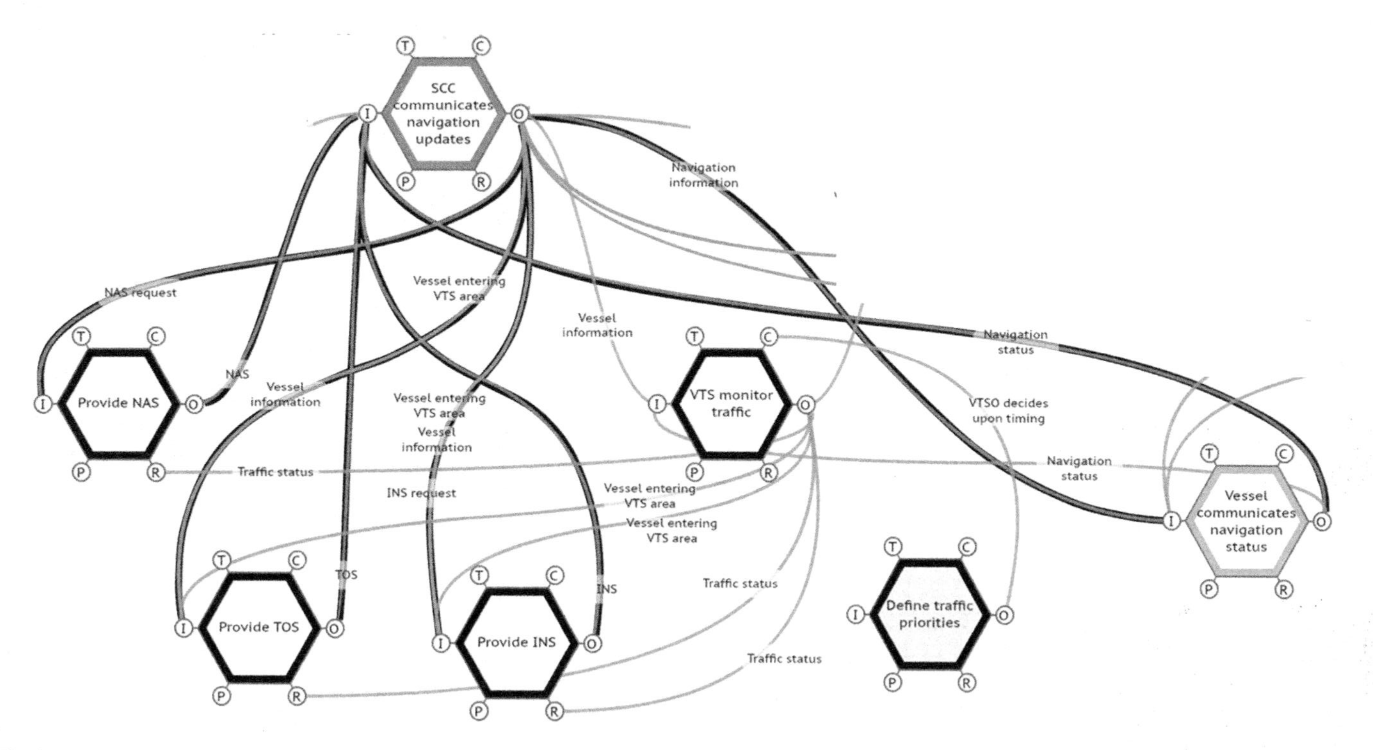

Fig. 2 Feedback loops around SCC communications

for the safe conduct of the vessel. Until that handover takes place, the responsibility for safe navigation will stay with the OOW in the SCC, which means that all requests from the autonomous vessel to the VTS will have to be transmitted through the SCC. SCC operators may be able to anticipate some or most of the information needs and ensure that VTS provides it before the crew aboard the vessels communicates its request, or the crew may also request information before it actually becomes necessary. However, while the whole control transfer process is ongoing, the vessel will keep navigating towards the port under remote control or autonomous mode and traffic around the port area will naturally also remain in full operation. This means that time pressure may easily become a critical factor and the ability to respond to any unanticipated events becomes quite unclear.

Figure 3 shows an instantiation of the FRAM model for what could be the control handover process, with a particular focus on the exchanges of information that are likely to be needed. The numbers in black indicate the sequence of activation of the functions that are likely to be directly involved in the handover process. The colour codes on the functions illustrate the amplitude of variability in the output of that given function. The colour at the top of the function represents an estimation of the potential variability in the output of that function with regard to time, whereas the colour at the bottom represents an estimation of that variability in terms of the precision of the output. Progressively darker red tones are used to indicate increasing amplitude of variability that is actually observed in the output of the function, and blue and green colours indicate lower amplitude of variability. Naturally, each output may assume many different degrees of variability, but for the purpose of this discussion, the instantiation in Figure 8.3 shows what could be considered a "worst case scenario", with particular focus on high amplitude variability in the output of the function "control handover", as this represents the operational goal of the system here modelled.

While this is a simplified overview of the process and other interactions are likely to be carried out, it illustrates how coordination may become a critical aspect of future operations and how the ability to respond to unforeseen circumstances may compromise the entire handover process. A more thorough analysis would be needed to detail all the potential issues that may arise during boarding and handover. Based on data currently available, the following ones were highlighted:

- The crew is delayed due to difficulties in boarding the vessel (i.e. weather or sea conditions).
- The systems check report failures that were not previously detected from remote control.
- Conditions aboard the vessel do not match what was expected by the crew, and adjustments have to be made to planned operations.
- Navigation information is not provided in a timely way, which may lead to the need to revise plans for navigating into port or even the voyage plans.

As this instantiation only presents one possible situation, the process may be adjustable to changes, and it may not have to be precisely carried out according to the sequence that is represented in Fig. 3. As suggested by the numbers, some

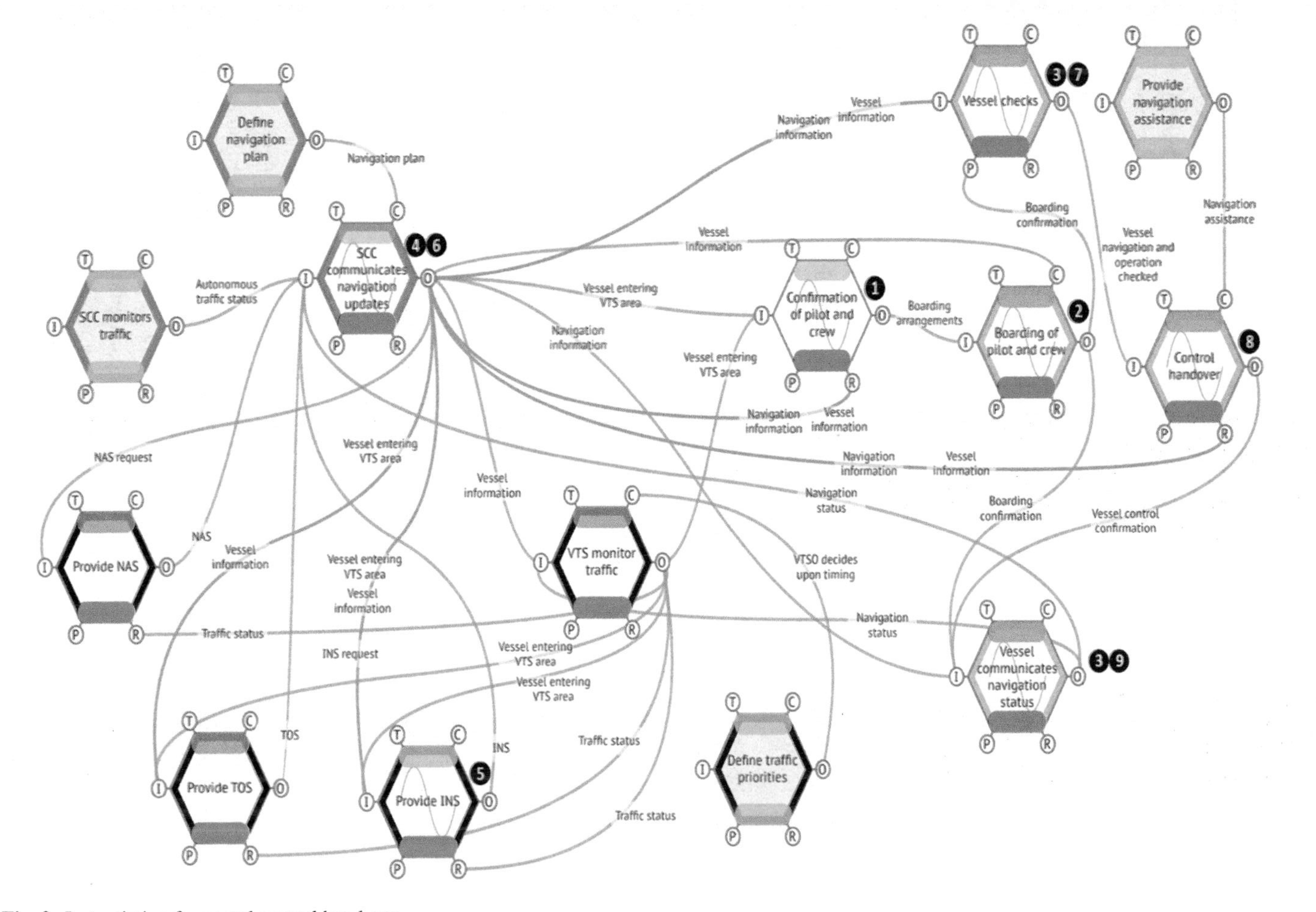

Fig. 3 Instantiation for vessel control handover

functions may even be carried out simultaneously. However, this will surely have repercussions in terms of the overall uncertainty and complexity that emerges from shifting control while the vessel keeps moving towards the port. Hence, the variability of functions to be carried out by the crew boarding the autonomous vessel becomes critical for whole system operation, particularly when other similar processes are likely to be ongoing simultaneously, which will increase uncertainty on SCC operations. Generating capacity for the crew onboard to adjust to unforeseen circumstances without compromising the safety of navigation into port also requires that the crew initiates boarding arrangements with much more anticipation that, for instance, currently pilots do. This will naturally impose additional resource constraints and may often be compromised by sea conditions.

If delayed boarding the vessels and aiming not to compromise port arrival schedule, the crew might attempt to compensate by expediting systems and vessel checks. This may in fact enable the control handover to be undertaken in such a way that the ship may continue navigating into port according to schedule. However, this means that in practice the crew may be operating and deciding based on more substantial and diverse assumptions (i.e. everything is OK to proceed with control handover). The output of the function "vessel checks" would become significantly delayed and/ or imprecise, and as illustrated in Fig. 3 by the colour codes, the variability of the output of most other functions may also be amplified, as to a certain extent, they rely on the precision and timing within which the crew boards the vessels and undertakes the necessary arrangements for control handover. The main potential impacts are shown in Fig. 3 through the waves in the functions.

6 Communication, Coordination, and Complexity

The modelling of anticipated functions in the future maritime transport system reveals many crucial aspects, which have up to now not been addressed in the literature. It also shows the potential complexity of introducing an increased degree of automation in vessel operations.

While some of the current service functions will remain largely unchanged, such as the role of the VTS overseeing and informing traffic with the goal to facilitate fluent, efficient, and safe traffic movements in and out of port, the preconditions are changed by the introduction of an additional service, the SCC that is likely to be operated by the shipping company/ies. Thus, an increased need to coordinate and communicate between SCC, conventional vessels, and VTS is anticipated. As traffic dynamics and complexity increase, so will the difficulties for VTS to acquire a suitable overview of navigation conditions in and out of port. This can be expected to significantly increase the exchanges of information, particularly as new control layers are added to the system. Communications might develop an iterative nature, as the need to confirm, verify and update traffic information becomes increasingly frequent and difficult.

Although not fully explored here, degraded operational modes are likely to raise many other related issues. In the case of systems' failures, particularly under the critical scenario of control handover previously explored, operators will have to take over the failing automated functions, and interactions between operators and systems that remain operational will intensify. The operational and safety requirements for automation under similar scenarios have been widely addressed in literature (Bainbridge 1983, Balfe et al. 2012) but nevertheless remain short of expectations.

The interactions between "conventional" vessels, those with advanced automation, and autonomous ones, can be expected to generate additional complexity issues. The coordination among the traffic participants is strongly reliant on the exchange of information. Particularly, for VTS services, there is the need to shift from data link (coming from vessels with automated systems) to voice communication protocols (from "convention" vessels), which is likely to raise complex challenges and an increased workload for the shore-based operators. Further, one of the core problems with shore control identified in the discussion is the aspect of command. If command is transferred from a shore centre to the ship, which is being manned by navigators, crew, and pilot, how is the takeover organized and how is accountability and responsibility for safety of navigation assigned. This will require new procedures with regard to the physical handover between shore and ship and deeper understanding of the roles of different actors.

The high dynamics and complexity aspects outlined are well within the scope of resilience engineering thinking (Nemeth & Hollnagel 2014). The highly distributed and opportunistic nature that maritime navigation currently retains seems compatible with the key principle of generating adaptive capacities (Woods 2015). However, the introduction of different control layers and operational concepts (i.e. centralised and automated systems) will inevitably transform the way navigation conflicts are currently negotiated between vessels. The growing congestion around major ports also erodes the buffering capacities that may have so far facilitated such negotiations. The persistence of collisions and groundings as major safety issues in the maritime domain provides evidence towards the stretching of capacity boundaries under current operation concepts, particularly around worldwide major ports that are already showing capacity problems.

7 Opportunities for Enhanced Resilience in Future Traffic Management

Currently, the maritime transport system can be understood as a loosely coupled complex system. The traffic largely acts independently, and the VTS oversees traffic flows and informs traffic if needed. Despite a limited capability for tactical and strategic control (Praetorius & Hollnagel, 2014), the system is rather well-adapted to the current operational settings with the ability to adapt and cope with quick contextual changes. However, through the anticipated changes with regard to the increased need for coordination and communication, the abilities to respond, monitor, anticipate, and learn will drastically be affected. While some coordination and communication requirements may be effectively formalised, namely, through operational procedures, resilience engineering literature has frequently argued the need for such elements in the scope of informal and flexible adjustment of work to local conditions. In the face of the foreseeable intensification of maritime traffic, particularly in the proximity of major international ports, these local conditions are likely to become increasingly specific and dynamic, which means that the need for informal coordination and communication also becomes more prominent. The FRAM-based approach in this chapter and its further exploration may pave the way towards developing an operational (functional) perspective on critical coordination and communication needs, as opposed to one purely based on the business-oriented needs that tend to focus more on the alignment of responsibilities and the formal roles within organizations. The insights developed through the FRAM can thus inform the design of future operations in such a way that coping with increased variability and uncertainty is better supported by flexible exchanges between actors in the system.

Responding characterizes a system's ability to know how to react in a given situation. This requires timely information about the system state as well as the possibility to act based on it. Within the anticipated system, a novel control layer is introduced through the SCC. This means partially the introduction of a centralised control feature that may reduce variability and, therefore, increase predictability of traffic movement. Thus, it can be argued that the ability to respond to development could potentially be enhanced in future operations.

Further, as the need for coordination between VTS, traffic, and SCC increases, the ability to monitor and anticipate will gain in importance. New indicators for safe operations need to be developed to be able to determine the system's current and potential future states to be able to prepare for and cope with both routine and irregular operations in a dynamic operational context. This will partially be possible through traditional risk assessments, but will also require to revise potential sets of indicators once SCC and autonomous traffic start to operate. It is important to take the effects of the increased complexity into concern, as these will impact on what indicators can be considered as representative for different system states. Indicators for performance, therefore, need to address both the process of traffic management in autonomous shipping and its potential outcome(s). It is common to assess maritime operations and safety within these by outcome indicators, such as number of incidents and accidents, or traffic density in an area. However, to ensure safety in operation within autonomous vessel settings will also require to ensure that essential buffers in terms or resources (time, manpower, technology, procedures) for deviations in normal operations and abnormal situations are secured to ensure that the traffic system can maintain its functioning and cope with these. Therefore, process and outcome indicators are needed. The above FRAM analysis can serve as a tool to highlight potential challenges and generate discussions on what issues should be taken into consideration, and how they should be approached in view of their wider system relations.

As the need for coordination and communication increases, so do time constraints to find appropriate responses for situations that occur on less regular basis. Thus, the introduction of SCC and autonomous traffic needs to consider how to identify potential safety and security threats early on, that is, how to maintain and probably enhance the ability to anticipate. This requires new strategies to forecast traffic developments, as well as a clear definition of roles among the actors so that uncertainties in responsibilities are reduced as much as possible. Anticipation will play a crucial role to ensure that appropriate indicators to monitor current operations can be developed, maintained, and revised as needed.

This chapter has analysed a system that is still under development. It is, therefore, hard to determined how the ability to learn can be addressed. To ensure learning from positive and negative operational experiences is important, but this requires a system to be in place and operational. However, as maritime operations tend to show a rather reactive way of learning, we emphasize that future developments need to take current everyday work, and challenges therein, into consideration when a novel control and organizational structure is developed. Otherwise, there is a risk that today's operational challenges will just be transferred into future operations in addition to whatever novel demands may arise.

8 Conclusion

This chapter has explored future maritime operations through the lens of FRAM. While most of the research up to now has primarily focused on the ship side, this has been an attempt to understand the consequences of change to maritime traffic management including shore-side services such as the VTS. The analysis has shown that the increased automation primarily affects the system's capabilities and characteristics related to cooperation and communication among ship and shore, especially between the anticipated SCC and VTS.

While many stakeholders currently emphasize the potential of autonomous shipping in terms of efficiency and safety, the analysis has shown that more attention should be paid to the increased complexity and functional dependencies that arise based on the introduction of the SCC. This will affect the amount of information available, as well as the ability of services, such as VTS, to be able to monitor, respond, and anticipate to developments in the area. Furthermore, through the introduction of the SCC functions, the overall control settings in the maritime transport systems are changed from distributed to polycentric control. This will have an effect on the resources, that is, time, communication systems or data streams, required to uphold fluent, efficient, and safe traffic movements within port approaches. As a well-functioning coordination between ship, shore and the VTS is the focal point to ensure safe navigation, it is necessary to secure that the resources needed can actually be deployed in the right time.

Within this chapter, FRAM has enabled the visualization of the complexity within the coordination and communication processes between VTS, vessel, and

SCC. However, this should only serve as a starting point for further exploration. Rationalising around the "four resilience cornerstones" (Hollnagel 2009) in combination with the FRAM models has provided a useful approach for a discussion on future operational and safety-related challenges. For both VTS and SCC functions, monitoring and anticipating will become increasingly relevant, as maritime traffic around ports becomes more complex and difficult to predict; thus more buffering capacity is required to maintain the ability to quickly respond and adapt to changes in the operational context. The FRAM model helps to visualise how the introduction of centralised control features can help to reduce variability, and therefore increase predictability. However, this will also have effects on the system's ability to quickly adapt to situations where operations may deviate from normal procedures. Beyond safety compliance needs and the demonstration of independent systems operability, the exploratory work presented here shows how the FRAM can provide the basis for a prospective analysis of future operation concepts, and support the identification of where the challenges of "working across boundaries" may emerge.

References

Bainbridge, L. (1983). Ironies of automation. *Automatica, 19*(6), 775–779.

Balfe, N., Wilson, J. R., Sharples, S., & Clarke, T. (2012). Development of design principles for automated systems in transport control. *Ergonomics, 55*(1), 37–54.

de Vries, L. (2017). Work as done? Understanding the practice of sociotechnical work in the maritime domain. *Journal of Cognitive Engineering and Decision Making, 11*(3), 270–295.

Dekker, S., Cilliers, P., & Hofmeyr, J. H. (2011). The complexity of failure: Implications of complexity theory for safety investigations. *Safety Science, 49*, 939–945.

Ferreira, P. N. P., & Cañas, J. J. (2019). Assessing operational impacts of automation using functional resonance analysis method. *Cognition, Technology & Work, 21*(3), 535–552.

Hollnagel, E. (2009). The four cornerstones of resilience engineering. In C. Nemeth, E. Hollnagel, & S. Dekker (Eds.), *Preparation and restoration. Resilience engineering perspectives* (Vol. 2, pp. 117–134). Ashgate.

Hollnagel, E. (2012). *FRAM, the functional resonance analysis method: Modelling complex socio-technical systems*. Ashgate.

Hult, C., Praetorius, G., & Sandberg, C. (2019). On the future of maritime transport: Discussing terminology and timeframes. *TransNav--International Journal on Marine Navigation and Safety of Sea Transportation, 13*, 269–273.

International Association of Marine Aids to Navigation and Lighthouse Authorities (IALA). (2016). *Vessel Traffic Service Manual.* (6th ed.). https://www.iala-aism.org/

International Maritime Organization. (2019). *Autonomous shipping.* http://www.imo.org/en/MediaCentre/HotTopics/Pages/Autonomous-shipping.aspx

Lambou, M.A., & O. Masaharu. (2017). Shipping 4.0: Technology stack and digital innovation challenges. *Proceedings of the International Association of Maritime Economists (IAME).* Kyoto, Japan.

Nemeth, C., & Hollnagel, E. (Eds.). (2014) *Becoming resilient.* Resilience Engineering in practice, (Vol. 2). Ashgate Publishing.

Praetorius, G., & Hollnagel, E. (2014). Control and resilience within the maritime traffic management domain. *Journal of Cognitive Engineering and Decision Making, 8*(4), 303–317.

Praetorius, G., Hollnagel, E., & Dahlman, J. (2015). Modelling vessel traffic service to understand resilience in everyday operations. *Reliability Engineering & System Safety, 141*, 10–21.

Relling, T., Lützhöft, M., Ostnes, R., & Hildre, H. P. (2018). A human perspective on maritime autonomy. In *Proceedings of the international conference on augmented cognition* (pp. 350–362).

van Westrenen, F., & Praetorius, G. (2014). Situation awareness and maritime traffic: Having awareness or being in control? *Theoretical Issues in Ergonomics Science, 15*, 161–180.

Wiener, E. L., & Curry, R. E. (1980). Flight-deck automation: Promises and problems. *Ergonomics, 23*(10), 995–1011.

Woods, D. (2015). Four concepts for resilience and the implications for the future of resilience engineering. *Reliability Engineering and System Safety, 141*, 5–9.

A Methodological Framework for Assessing and Improving the Capacity to Respond to the Diversity of Situations That May Arise

Eric Rigaud

Contents

The capacity to respond to the diversity of situations that may arise is one of the cornerstones of safety management's Resilience Engineering perspective. This chapter focuses on the description of a framework aiming to collect and analyze data for supporting its assessment and the proposal of corrective actions. Resilience Engineering's theoretical background endorses the definition of performance indicators. Individual and collective interviews help the identification of factors to be corrected and others to be preserved.

E. Rigaud (✉)
MINES Paris Tech, PSL-Research University, CRC, Paris, France
e-mail: eric.rigaud@mines-paristech.fr

C. P. Nemeth, E. Hollnagel (eds.), *Advancing Resilient Performance*,
https://doi.org/10.1007/978-3-030-74689-6_9

1 Introduction

The capacity to respond to the diversity of situations that may occur is one of the critical cornerstones of resilience engineering (Hollnagel, 2011). To be considered resilient when facing an abnormal condition, agents have to adjust their behavior to prevent unwanted outcomes and continue accomplishing their duties according to their model of performance. Depending on the situation's nature, agents adapt their behavior by considering their experience, rules and procedures, leadership, or improvisation.

A review of the development of resilience metrics in the railway domain (Besinovic, 2020) demonstrates that resilience metrics are developed to support the network's robustness to disturbances and support the optimization of the train schedule and reschedule. Ferreira (2011) applied the Resilience Engineering perspective to railway planning activities experiments in the Resilience Engineering domain. Siegel and Schraagen (2014) propose a so-called resilience state model for railway systems adapted from Rasmussen's (1997) system boundaries and Woods and Wreathall's (2008) stress-strain model. De Regt, Siegel, and Schraagen (2016) propose metrics to quantify weak resilience signals.

The framework proposed in this chapter focuses on the sociotechnical system's capacity to respond and aims to support its formalization and its assessment by identifying essential factors to be preserved and vulnerability factors to be corrected. The first section describes the theoretical background shaping the development and the different phases that structure the framework's application. The following sections detail them. Finally, the last part describes a synthesis of the results of its implementation.

2 Rationale for the Overall Approach

The framework helps identify and handle gaps and needs related to a system's ability to respond to the diversity of situations that may arise in a systematic and structured manner (Rigaud et al. 2013, 2018). Safety managers can use the framework to enhance their understanding of the system's complexity and structure learning, training, and change management activities to improve their operations' security. They can apply it at a different scale (technological system, process, unit, plant).

3 Basis and Sources of the Framework

The Resilience Engineering perspective on safety management structures the framework. Borys, Else, and Leggett (2009) consider Resilience Engineering as the fifth age of safety. This period follows a phase of integration (Glendon et al., 2006) of technical, human, managerial, and cultural factors in risk management practices (Hale & Hovden, 1998).

Douglas and Wildavsky (1983) consider that no one can know and predict all the potential risks and associated consequences. Risks are selected using rational and irrational criteria. However, even within the scientific community, there is rarely a consensus regarding potential risks and accompanying problems. The Resilience Engineering perspective aims to endow systems with the requisite imagination to respond and overcome the diversity of situations that can occur (Adamski & Westrum, 2003, Woods & Hollnagel, 2006). The aim is to change the main focus of safety management from the prevention of risks to the development of workers' adaptive capacity to be in control despite the variability and the complexity of situations and the lack of time, knowledge, competence, or resources (Hollnagel & Woods, 2006). The target is the development of the resilience of systems. Resilience refers to the "intrinsic ability of a system to adjust its functioning before, during, or following changes and disturbances, so that it can sustain required operations under both expected and unexpected situations" (Hollnagel, 2011). It also refers to the "ability to recognize and adapt to handle unanticipated perturbations that call into question the model of competence, and demand a shift of processes, strategies, and coordination" (Woods, 2006).

The capacity to respond to regular and irregular variability, disturbances, and opportunities either by adjusting the way things are done or activating readymade responses is one of the four essential capacities that structure the conceptualization of system resilience (Hollnagel, 2011). The three others are the capacity to monitor changes, the capacity to anticipate developments, threats, and opportunities, and the ability to learn the right lessons from the right experience.

4 Theoretical Background

The framework's theoretical background is composed of seven situations aiming at describing the diversity of conditions that can occur within the system for supporting data collection and of two performance indicators aiming at supporting the assessment of the system performance.

5 Situations of Resilience

Assessing the capacity to respond following the Resilience Engineering perspective on safety management requires considering different situations and associated adaptive behavior. Five variables structure the definition of these situations:

- *The type of adverse situations.* Firstly, the adverse situation classification considers if the system finds them normal or abnormal, and secondly, their predictability.

Thus, the typology considers four types: normal situation, regular abnormal situation, abnormal irregular situation, and exceptional/unexampled situation.

- *Adaptive processes.* The functions considered for describing the adaptive processes aiming at responding to the different situations are: 1) event detection, 2) situation recognition, 3) decision to act, 4) definition of the behavior, 5) mobilization of resources, 6) act.
- *Existence of good practices and/or procedures.* For each adaptive process identified, the existence of good practices and/or procedures is considered.
- *Context of action.* The context of action is related to the difference between competence, knowledge, resources, and time required to perform adequately adaptive processes identified and the competence, knowledge, resources, and time available.
- *Performance model.* The criteria used for assessing the system performance are quality, reliability, safety, security, sustainability, etc.

The variables induce seven situations of resilience to consider when collecting data and assessing the capacity to respond:

1. The situation is normal, considered by procedure or good practices, and the context (time, knowledge, competencies, and information) necessary to respond is available. Agents can recognize the situation, define their future behavior by using their experience or by adapting a known and regularly applied procedure, and apply it in conformity with all the dimensions of performance of the activity.
2. The situation is normal, considered by procedure or good practices. However, the context (time, knowledge, competencies, information) necessary to respond is not available. Agents can recognize the situation, define their future behavior by using their experience or adapting a known and regularly applied procedure and apply it with creativity to conform with all dimensions of the activity's performance despite the lack of one kind of resource.
3. The situation is normal and not considered by procedure or good practices. Agents can recognize the situation and that neither procedure nor good practices support them to define the behavior to adopt, they are creative to define their future behavior and apply it in conformity with all dimensions of performance of the activity.
4. The situation is abnormal (perturbation, crisis), considered by procedure or good practices, and the context (time, knowledge, competencies, and information) necessary to respond is available. Agents can recognize the situation and the necessity to adopt a non-routine behavior; they define their future behavior by using their experience or by adapting a known procedure or find one in a guideline, they apply it in conformity with all the dimensions of performance of the activity in contributing to the continuity of the activity of the system.
5. The situation is abnormal, considered by procedure or good practices, but the context (time, knowledge, competencies, information) necessary to respond is not available. Agents can recognize the situation and the necessity to adopt a non-routine behavior; they define their future behavior by using their experience or by adapting a known procedure or find one in a guideline, and apply it with

creativity in order to conform with all dimension of performance of the activity despite the lack of one kind of resources.
6. The situation is abnormal and not considered by procedure or good practices. Agents can recognize the situation and the necessity to adopt a non-routine behavior. Neither procedure nor good practices support them to define the behavior to adopt. They are creative in defining their future behavior and apply it in conformity with all dimensions of the activity's performance.
7. The situation is unexampled. Agents are creative to respond and to contribute to the continuity of activity of the system.

6 Performance Indicators

The seven situations will support data collection and system description. Two performance indicators support the evaluation of the system's capacity to respond.

The first indicator is related to the capacity of operational agents to adjust their procedural or methodological framework or be creative to carry out their regular activity despite the variability of their environment while respecting the temporal, economic, and activity-specific performance criteria. Four rules structure the indicator:

1. Agents know their work and associated performance criteria.
2. They have the skills or know the procedures to follow and have the resources, time, and information to follow the different performance criteria.
3. If they lack skills, resources, time, or information, they can be creative in carrying out their work according to performance criteria.
4. If the situation changes and the procedural framework is no longer applicable, they can be creative enough to carry out their work following performance criteria and have the necessary maneuver margins.

The second indicator is related to the capacity of operational agents to adjust their normative or methodological framework or to be creative in order to face and overcome the occurrence of an urgent or unexpected situation, anticipated or not while respecting the temporal, economic, and activity-specific performance criteria.

The four rules associated with the indicator are:

1. Agents are aware of the abnormal situations, the behavior to adopt when they occur, or what document to consult.
2. They have the skills, resources, time, and information to respond to the situation following the different performance dimensions.
3. If they lack skills, resources, time, or information, they can be creative in responding to the situation following the different performance dimensions.
4. If the situation changes and the procedural framework is no longer applicable, or there is no procedural framework, they can be creative in responding to the situation.

This conceptual background supports the application of the framework. The following section describes the different phases to follow for conducting the assessment.

7 Key Elements of the Framework

The framework's application consists of conducting workshops, individual interviews, focus groups, and observations for collecting, analyzing, and presenting data related to the performance of resilience, factors to be preserved and corrected, and action plans for developing resilience. The framework is a modular system of different elements (Module 1–4). The complete process model is only needed in case of the first implementation in the organization. The framework consists of four modules:

1. *Definition of the context of the study.* The first step involves clearly defining the goal and scope of the study. The team organized workshops for describing the system studied, the diversity of events it has to respond to, and its capacity to respond. An assessment methodology and associated supportive material (diagnostic schedule, interviews and observation guidelines, assessment grid) are derivates from this context.
2. *Data collection.* The second step aims at collecting data related to the system's capacity to respond. The team conducts individual and collective interviews and observations for collecting qualitative and quantitative data about the system structure and dynamic in regular times and when disturbances occur by considering the different actors of the system (operational, managers, and directors).
3. *Diagnosis.* The third step consists of analyzing data collected to provide a resilience score and a list of factors to be preserved and corrected.
4. *Definition of an action plan.* The fourth step consists of providing a set of actions to develop resilience by correcting negative factors and highlighting and preserving positive factors.

8 Roles and Responsibilities

A set of essential roles supports the distribution of responsibilities when applying the method, considering that one person can assume different roles.

- The "evaluation owner" is the person who is mainly responsible for the system to be assessed. This critical role encompasses the following responsibilities: defining the goal and scope of the evaluation process, supporting the assessment team in providing access to the agents of the system, and to document resources needed by the assessment (room, material, etc.).

- The "evaluation coordinator" is the person who is mainly responsible for the evaluation process. The evaluation coordinator should cover the following responsibilities: defining the target, the scope, and the objective of the evaluation process with the "evaluation owner," planning the different steps of the assessment, monitoring the realization of the different steps, managing issues when performing the different steps.
- The "stakeholder coordinator" is the person who is mainly responsible for the coordination with the various agents involved in the assessment. The stakeholder coordinator should cover the following responsibilities: identifying the agents, invite the agents to workshops, provide feedbacks of the assessment to the agents.
- The "technical coordinator" is the person who is mainly responsible for the realization of the assessment task. The evaluation coordinator should cover the following responsibilities: organizing and animating workshops, writing deliverables.

The following sections describe the rationale of the four modules. They describe the objective of the phase and practical information to conduct associated workshops.

8.1 Data Sought and Reason(S) for Choosing

The first step aims at defining the context of the assessment process. It involves: (1) defining the goal and scope of the study, (2) describing the resilience of the system assessed, (3) organizing the assessment. The team in charge of the assessment organizes workshops for achieving these tasks. The following section presents the different workshops.

Defining the goal and scope of the study : The first task to achieve consists in the definition of the general context of the assessment process. The "evaluation owner" and the "evaluation coordinator" define the goal and the scope of the study (cf. Table 1)

Describing the system resilience : The second task to achieve consists in the description of the system to be studied and its associated capacity to respond to performance. The "evaluation coordinator" assisted by the "technical coordinator" describes the system and defines the capacity to respond to capacity performance by considering the context of the study (cf. Table 2).

Organizing the assessment : The third task to achieve consists of the definition of the assessment project by planning the assessment, assigning roles and responsibilities, and designing material (cf. Table 3)

After achieving the three steps, all the elements to conduct the data collection and the assessment are available.

Table 1 Goal and scope definition task

Definition of the context of the study Goal and scope definition			
Target To define the goal and the scope of the study	Duration 2 * 2 h	People needed Evaluation owner Evaluation coordinator	
In a nutshell	The evaluation owner nominated the coordinator of the evaluation. They together define the system to be studied, the results intended after the assessment, and the preliminary calendar.		
Methods Brainstorming, discussion	**Tools** Methodological guideline	**Input** Methodological guideline, generic indicators	**Output** Goal, scope, and preliminary calendar
In depth	The application of the framework can cover the whole system or only selected parts. The selection of topics addressed by the study will affect the process, the working team, the duration of the study, and the stakeholders involved. The evaluation owner launching the study has to nominate a coordinator, and together they have to predefine: ■ The system studied. It can be a process, a plant, a workspace, a task, etc. ■ The goal. It can be prospective (to test the resilience engineering perspective) or specific (benchmarking, challenge a procedure, a process, a continuity plan, etc.) ■ The calendar. Three months is necessary to perform different tasks. ■ Stakeholders. A preliminary list of agents to involve within the study. ■ The technical and the stakeholder coordinator identity.		
Checklist	1. Evaluation coordinator nominated. 2. System to studied defined. 3. Objective of the assessment defined. 4. Preliminary calendar defined. 5. Preliminary stakeholders list defined. 6. Technical coordinator nominated. 7. Stakeholder coordinator nominated.		

8.2 Data Collection

The second step aims at collecting the data required to proceed with the performance assessment. It involves: (1) presenting the assessment context, aim and methodology and (2) collecting data.

Present the assessment context, aim and methodology : The first task to achieve consists of explaining to the stakeholder the context, the objective, and the organization of the assessment (cf. Table 4)

Collect data : The data collection process aims at collecting information required to proceed with the performance assessment (cf. Table 5)

Table 2 System resilience description task

Definition of the context of the study System resilience description			
Target To describe the system and its associated capacity to respond to performance.	Duration 6 * 2 h	People needed Evaluation coordinator Technical coordinator	
In a nutshell	The evaluation coordinator and the technical coordinator propose a sociotechnical description of the system studied and of its capacity to respond.		
Methods Brainstorming, system modelling	**Tools** Methodological guideline Situation of resilience Performance indicators	**Input** Target, goal and scope of the study	**Output** Description of the system and its capacity to respond
In depth	The definition of the system targeted by the study is a critical phase. The evaluation and the technical coordinators describe the system by considering the relationships between goals, processes, procedures, people, building, infrastructure, technology, and culture at the scale of the sociotechnical system and the influence of external factors such as economic circumstances, regulatory frameworks, and stakeholders. They gather data from appropriate sources, including key actors, stakeholders, subject-matter experts, and internal and external documents for describing each dimension of the system and their interactions. Then they define the capacity to respond by considering the typology of situations of resilience and the performance indicators. They synthesized the results in a document that will base the data collection and assessment processes.		
Checklist	1. Description of the system. 2. Description of the capacity to respond.		

Data analyses The third step consists in analyzing data collected to provide a resilience score and a list of factors to be preserved and to be corrected. It involves (1) indicator's evaluation, (2) formalization of factors of resilience and vulnerability, and, (3) writing of the preliminary report.

Indicator's evaluation: When the assessment team achieved the data collection phase, they proceed to the analysis of data. This process consists of rating two indicators with the support of the data collected (Table 6).

Formalization of factors of resilience and vulnerability: Besides, the assessment team formalizes two lists of factors. The first list is labelled "resilience factor"; the second list is named "vulnerability factor." (Table 7)

Table 3 Organizing the assessment task

Definition of the context of the study Organizing the assessment			
Target To define the system resilience assessment project.	Duration 3* 2 h	People needed Evaluation owner Evaluation, technical and stakeholder coordinators	
In a nutshell	The team designs the project aiming at assessing the capacity of response of the system with planning data collection, assessment and results, discussion tasks, and assigning roles and responsibilities.		
Methods Brainstorming, process modelling, project management	**Tools** Methodological guideline, flow diagram	**Input** Assessment objectives, system description	**Output** Assessment project
In depth	The third step consists of finalizing the definition of the context of the assessment by formalizing the associated project. The team answers the following questions: ▪ Which data are needed to assess the performance indicators? ▪ Who will collect the data, when, and how? ▪ How much time and resources are available? ▪ Which material is required to collect data? The report describing the context of assessment contains a short and precise definition of the assessment objective, a description of the methodology and all management information to be known with the relevant stakeholders of the project (issues and objectives for each actor, actor's description, organization chart, project management, assessment planning and tasks, communication plan, meetings, performance indicators).		
Checklist	1. Data to be collected 2. Data collection methodology 3. Assessment project 4. Roles and responsibilities 5. Data collection and analysis material		

8.3 Writing and Validation of the Resulting Report

The third task of this phase consist in writing a preliminary version of an assessment report and validating it with stakeholders (Table 8)

8.4 Recommendations and Action Plans

The fourth step consists of providing a set of actions to be performed to develop the performance of resilience by correcting negative factors and highlighting and preserving positive factors. It involves (1) hierarchization of the resilience and vulnerability factors and (2) actions identification.

Table 4 Present the assessment context, aim, and methodology task

Data collection Present the assessment context, aim, and methodology			
Target To explain the stakeholder, the essential information related to the assessment	Duration 2 h	People needed Evaluation, technical and stakeholder coordinators	
In a nutshell	The team initiates the stakeholders to the resilience perspective on safety management and to the context and organization of the study		
Methods Presentation	**Tools** Methodological guidelines	**Input** Assessment context	**Output** Stakeholders informed
In depth	The resilience engineering perspective on safety management is a new approach. Consequently, the first step of the data collection consists of presenting resilience engineering to the stakeholder and the essential information about the assessment of the stakeholder. The presentation describes: ▪ The resilience engineering perspective on safety ▪ Objectives of the assessment ▪ Assessment methodology ▪ Assessment schedule At the end of the workshop, stakeholders should have understood the context of the assessment and their future contributions.		
Checklist	Stakeholders informed about the assessment		

Hierarchization of the resilience and vulnerability factors: The first task of this phase consists in ranking resilience and vulnerability factors (Table 9).

Identification of actions: The second task consists in defining short- and long-term actions for preserving resilience factors and correcting vulnerability factors (Table 10).

Finally, the assessment team provides the final report presenting the system studied, the methodology followed, the results of the assessment, and the action plan defined.

9 Lessons from the Application of the Framework

This section presents the lessons of the application of the framework in the railway industry. The evaluation owner was responsible for the station's train traffic. The coordinator was a railway expert in human factors. The stakeholder coordinator was the manager of train departure/arrival processes, and the technical coordinator was an expert in resilience engineering. This team collaborates to accomplish the four phases of the study.

Table 5 Collect data task

Data collection Collect data			
Target Collect data required to assess the performance of the system	Duration 10*2 h	People needed Technical and stakeholder coordinators Stakeholders	
In a nutshell	The team collects data required to perform the assessment by observing stakeholders accomplishing their task and by conducting interviews.		
Methods Observation, interviews	**Tools** Methodological guidelines	**Input** Assessment project	**Output** Data collected
In depth	The data collection process aims at collecting information required to proceed with the performance assessment. The team can use different methods for data collection: Document analysis, individual or collective interviews, questionnaires, or observation. The technical and stakeholder coordinator conduct: ■ Observations of the behavior of stakeholders when performing their tasks. ■ Individual and collective interviews about the resilience of the system and of the capacity to respond. ■ If necessary, a focus group can be organized to precise some topics. At the end of the process, all data required to assess the system is collected.		
Checklist	1. List of data collection tasks 2. Interviews of operational agents 3. Interviews of managerial agents 4. Interviews of safety managers		

Definition of the context of the study: The motivation of the study was to experiment with a resilience engineering-based assessment to identify the added value related to human factor assessment. The system studied was train departure and arrival processes. These tasks involve operational agents, first-line managers, and safety managers. Schedule, procedures, and time constraints structure the process. Injuries may occur, and events happening in the station, and in the network, might affect its functioning.

After a set of preliminary interviews, the team adapts the generic performance indicators. It produces a questionnaire aiming at collecting qualitative data aiming to describe the diversity and the complexity of the capacity to respond to the unwanted situation of the train departure and arrival processes.

Data collection: The technical coordinator interviews eight operational agents, six-team leaders' representatives of the different tasks of the departure/arrival process, the head of the safety management department, and the head of traffic management in the station, with a specific questionnaire. He spends one day observing the realization of the different tasks, and, with a human factor expert, they assist in a crisis management exercise.

Table 6 Assess indicators task

Data analyses Assessment of the capacity to respond			
Target To assess performance indicators.	Duration 2 * 2 h	People needed All the team	
In a nutshell	The team analyzed data collected in order to evaluate the two performance indicators associated with the capacity to respond.		
Methods Meeting	**Tools** Brainstorming	**Input** Data collected	**Output** Indicators assessed
In depth	This process consists of rating two indicators with the support of the data collected. The assessment team evaluates each rule by asking, "do the data collected allow answering true to the rule?". If there is a disagreement between the members of the team, they formalize the causes of the disagreement. Then, they collect a complement of information in order to be able to evaluate the rule.		
Checklist	1. Two indicators assessed.		

Table 7 Formalization of factors of resilience and vulnerability task

Data analyses Formalization of factors of resilience and vulnerability			
Target To define factors of resilience and vulnerability	Duration 2 * 2 h	People needed All the team	
In a nutshell	The team analyzed data collected in order to identify factors of resilience and factors of vulnerability.		
Methods Meeting	**Tools** Qualitative analysis Brainstorming	**Input** Data collected	**Output** Factors of resilience and vulnerability
In depth	The team analyzed data collected and produce two lists of factors: ▪ Resilience factors are properties allowing or promoting the capacity to respond. The team defines them in order that the system can sustain them when change happens and enhance them if possible. ▪ Vulnerability factors are barriers against an adequate capacity to respond. The team formalized them in order to correct them.		
Checklist	1. List of resilience factors 2. List of vulnerability factors		

Diagnosis: The first indicator provides insight into operational agents' adaptive capacity and the margin of maneuver provided by the system for overcoming the variability of routine situations. Operational agents demonstrate a good knowledge about the complexity of their tasks and find trade-offs between the different performance dimensions. They consider beginners' training and fear management,

Table 8 Report writing and validation task

Data analyses Writing and validating the resulting report			
Target To produce a preliminary report	Duration 6 * 2 h	People needed All the team stakeholders	
In a nutshell	The team writes a preliminary version of the report presenting assessment results.		
Methods Workshops	**Tools**	**Input** Preliminary results	**Output** Preliminary report
In depth	The team produces a preliminary report containing the following information: ■ Synthesis of the content of the report. ■ Description of the context of the framework methodology and assessment (aims, organization, working team). ■ Description of the data collection process. ■ Description of preliminary results of the assessment (indicators, resilience, and vulnerability factors). The team discusses the report during a validation workshop with stakeholders. During this workshop, they present and discuss the performance indicators values. If stakeholders disagree with some results, the team starts a debate in order to understand the issues and considers them in order to refine the results or conducts an additional investigation.		
Checklist	1. Indicator's evaluation validation 2. List of resilience factors validation 3. List of vulnerability factors validation		

Table 9 Resilience and vulnerability factors task

Recommendations and actions plan Hierarchization of the resilience and vulnerability factors			
Target To rank vulnerability and resilience factors	Duration 2* 2 h	People needed All the team stakeholders	
In a nutshell	Stakeholders rank resilience and vulnerability factors		
Methods Brainstorming	**Tools**	**Input** List of resilience and vulnerability factors	**Output** List of ranked resilience and vulnerability factors
In depth	The team organizes a workshop dedicated to the presentation and the ranking of resilience and vulnerability factors. Stakeholders discuss and establish a hierarchy by answering the following questions: ■ What are the most important resilience factors to be preserved? ■ What are the most important vulnerability factors to be corrected?		
Checklist	1. Ranked list of resilience factors 2. Ranked list of vulnerability factors		

Table 10 Action's identification task

Recommendations and actions plan Identification of actions			
Target To identify actions	Duration 2* 2 h	People needed All the team stakeholders	
In a nutshell	Stakeholders identify short- and long-term actions for preserving resilience factors and correcting vulnerability factors		
Methods Brainstorming	**Tools**	**Input** List of ranked resilience and vulnerability factors	**Output** Actions plan
In depth	The team organizes a workshop dedicated to the identification of actions plan for preserving resilience factors and correcting vulnerability factors. The brainstorming aims at identifying actions and ranked them by considering four types of actions: ▪ Short term and easy to implement ▪ Short term and complicated to implement ▪ Long term and easy to implement ▪ Long term and complicated to implement Stakeholders rank actions, and for the one considered as a priority, they provide content for supporting their practical realization (objectives, tasks, responsibilities, resources, schedule, criteria of success).		
Checklist	1. Actions plan for preserving resilience factors 2. Actions plan for correcting vulnerability factors		

procedures modification, tasks risks perception, and technological failures as sources of disturbances. They acknowledge having sufficient temporal margins of manoeuvre but have to compensate for human and technical resources' unavailability with increasing communication and coordination. Communication is an essential dimension of performance. The agent's objective is to deliver the right message to the right person at the right time. They use informal communication networks and personal information tools to complete the formal communication system. Many situations, incidents, delays, and malfunctions require an adaptive response. Managers have abilities to compensate for the absence of operational agents in performing their tasks. Agents take the initiatives to perform tasks. The hierarchy provides temporal margins, even if it creates some delays in the finalization of the process. Nevertheless, they require operational agents to follow procedures.

The second indicator addresses operational agents' adaptive capacity and the margins of manoeuvre provided by the system for overcoming abnormal situations such as incidents or accidents. Agents distinguished four types of abnormal situations: increased workload, safety incident into the station, crisis managed by the station, crisis managed by an authority external of the station. These situations induce increased tasks to achieve verbal and physical aggression, stress, unavailability of resources, difficulty or impossibility to apply procedures, or leadership and authority constraints. A culture of mutual assistance between agents and the "pride of the railwayman" contribute to agents' efforts for adaptation required to overcome disturbances.

Definition of an action plan: The technical coordinator presents the assessment results to the agent interviewed and representative of the train arrival/departure processes. For the two indicators, vulnerability and resilience factors identified were presented, illustrated, and discussed. The five essential resilience and vulnerability factors were selected. For each of them, brainstorming was conducted to identify short- and long-term changes to prevent vulnerability factors and preserve resilience factors. Solutions emerge; nevertheless, the absence of an available budget makes their application difficult. One feasible solution identified is the integration of the resilience engineering issues within human factors training already planned.

10 Conclusion

This chapter presents a methodological framework dedicated to assessing the capacity to respond to the diversity of situations that may affect a sociotechnical system. This framework uses traditional qualitative data collection methods and the theoretical background of resilience engineering.

The application of the method allows the identification of a set of lessons:

1. Agents are willing to discuss how they adapt when disturbances happen.
2. Talking about actions at the limit or outside the procedural context is complex with the hierarchy.
3. Budget optimization policies make difficult the realization of changes aimed at resolving vulnerability factors and preserving resilience factors.

The following steps in developing and validating the framework consist of applying it to another sociotechnical system and adapting it to consider cities' resilience.

References

Adamski, A., & Westrum, R. (2003). The fine art of anticipating what might go wrong. In E. Hollnagel (Ed.), *Handbook of cognitive task design*. Lawrence Erlbaum Associates.

Bešinović, N. (2020). Resilience in railway transport systems: A literature review and research agenda. *Transport Reviews, 40*(4), 457–478.

Borys, D., Else, D., & Leggett, S. (2009). The fifth age of safety: The adaptive age? *Journal of Health & Safety Research & Practice, 1*(1), 19–27.

Chandler, D. (2014). *Resilience: The governance of complexity. Critical issues in global politics*. Routledge.

De Regt, A., Siegel, A. W., & Schraagen, J. M. (2016). Toward quantifying metrics for rail-system resilience: Identification and analysis of performance weak resilience signals. *Cognition, Technology and Work, 18*(2), 319–331.

Douglas, M., & Wildavsky, A. (1983). *Risk and culture: An essay on the selection of technological and environmental dangers*. University of California Press.

Ferreira, P., Clarke, T., Wilson, J. R., et al. (2011). Resilience in rail engineering work. In E. Hollnagel, J. Paries, D. D. Woods, & J. Wreathall (Eds.), *Resilience in practice* (pp. 145–156). Ashgate Publishing.

Glendon, A., Clarke, S., & McKenna, E. (2006). *Human safety and risk management* (2nd ed.). CRC Press.

Hale, A. R., & Hovden, J. (1998). Management and culture: The third age of safety. A review of approaches to organizational aspects of safety, health and environment. In A. M. Feyer & A. Williamson (Eds.), *Occupational injury: Risk prevention and intervention*. Taylor and Francis.

Hollnagel, E. (2011). RAG: The resilience analysis grid. In E. Hollnagel, J. Pariès, D. Woods, & J. Wreathall (Eds.), *Resilience engineering in practice: A guidebook*. Routledge.

Hollnagel, E., & Woods, D. D. (2006). Epilogue: Resilience engineering precepts. In E. Hollnagel, D. Woods, & N. Leveson (Eds.), *Resilience engineering: Concepts and precepts*. Ashgate.

Rasmussen, J. (1997). Risk management in a dynamic society: A modelling problem. *Safety Science, 27*(2-3), 183–213.

Rigaud E., Neveu C., Duvenci-Langa S., Obrist M.-N., & Rigaud S. (2013). Proposition of an organisational resilience assessment framework dedicated to railway traffic management. In N. Dadashi, A. Scott, J. R. Wilson, & A. Mills (Eds.), *Rail human factors: supporting reliability, safety and cost reduction*. Taylor & Francis.

Rigaud E., Neveu C., & Duvenci-Langa S. (2018). Lessons from the application of a resilience engineering based assessment method to evaluate the resilience of a train departure and arrival management system. In S. Haugen, A. Barros, C. van Gulijk, T. Kongsvik, & J.-E. Vinnem (Eds.), *Safety and reliability – safe societies in a changing world*. CRC Press.

Siegel, A. W., & Schraagen, J. M. C. (2014). Measuring workload weak resilience signals at a rail control post. *IIE Transactions on Occupational Ergonomics and Human Factors, 2*(3-4), 179–193.

Woods, D. D. (2006). Essential characteristics of resilience. In E. Hollnagel, D. Woods, & N. Leveson (Eds.), *Resilience engineering: Concepts and precepts*. Ashgate.

Woods, D. D., & Hollnagel, E. (2006). Prologue: Resilience engineering concepts. In E. Hollnagel, D. Woods, & N. Leveson (Eds.), *Resilience engineering: Concepts and precepts*. Ashgate.

Woods, D. D., & Wreathall, J. (2008). Stress–strain plots as a basis for assessing system resilience. In E. Hollnagel, C. Nemeth, & S. Dekker (Eds.), *Remaining sensitive to the possibility of failure. Resilience engineering perspectives* (Vol. 1, pp. 145–161). Ashgate.

Addressing Structural Secrecy as a Way of Nurturing Resilient Performance

Alexander Cedergren and Henrik Hassel

Contents

By now, it is probably clear to most safety scholars that the concept of resilience has gained an incredible increase in popularity, with a plethora of different meanings attached to it (see Woods, 2015). The concept is used in as disparate fields such as engineering, sociology, and psychology (Alexander, 2013; Birkland & Waterman, 2009; de Bruijne et al., 2010; Pendall et al., 2010), and as pointed out by Boin et al. (2010), it almost appears that everyone and everything can, and should, be resilient.

Resilience engineering (RE) has drawn inspiration from several other research traditions (see Woods, 2003), and has offered a refreshing perspective in the domain of safety. When RE was introduced in 2006, it was rather confidently announced as a new "paradigm" in system safety research (Hollnagel & Woods, 2006). By giving special consideration not only to why things sometimes go wrong, but also why they usually go right (Hollnagel, 2008), RE has contributed to redefining the system safety discourse (although its novelty has been debated, see Haavik et al., 2019; Hopkins, 2014). In RE, performance variability is seen as a normal and necessary part of modern complex systems, and the adaptability and flexibility of human work to meet this variability is the driver behind effective and successful performance (Hollnagel, 2006). Sometimes, however, this intrinsic variability also leads to failure. In this way, it is the same mechanism creating both success and failure; they simply represent two sides of the same coin (Hollnagel, 2006). Without falling into the trap of treating success and failure in a simplified, binary manner, these insights

---

A. Cedergren (✉) · H. Hassel
Lund University, Lund, Sweden
e-mail: alexander.cedergren@risk.lth.se

C. P. Nemeth, E. Hollnagel (eds.), *Advancing Resilient Performance*,
https://doi.org/10.1007/978-3-030-74689-6_10

leave us with the conclusion that, in order to understand failure, we must also understand success (Woods & Hollnagel, 2006). In particular, we need to understand the way success is created through the continuous processes of adjustment and adaptation of people in the system of interest.

In literature on RE, resilience has been defined as the ability of a system to effectively "adjust its functioning prior to, during, or following changes and disturbances, so that it can continue to perform as required after a disruption or a major mishap, and in the presence of continuous stresses" (Hollnagel, 2009: p. 117). Significant attention in contributions to the field of RE has addressed conceptual ideas and expressions about how teams and organisations deal with complexity. Theoretical abstractions about the essential hallmarks of resilient systems, such as the abilities to anticipate, monitor, respond, and learn, have been postulated (Hollnagel, 2009). Moreover, theoretically thrilling concepts have been suggested as an attempt to capture how teams and organisations exhibit abilities to stretch in the face of surprises through, for example, processes of "graceful extensibility" (Woods, 2015). While such expressions may be praised in the world of academics, these contributions are difficult to translate into useful tools in the practical world. One of our main concerns for the future for RE is, therefore, the need to increase the extent of practical implementation of these inspiring conceptual contributions that have been introduced during the last decade. Given the practice-oriented focus of this book, our ambition is to share insights and work practices we have tried to increase the chances of, and foster more resilient performance in a public sector organisation with the goal of advancing RE not only as an academically stimulating "think tank" among scholars but also as a practically meaningful guidepost.

1 Background

This chapter draws from a three-year researcher-practitioner collaboration in the context of a public sector organisation. The specific case used as a point of departure for this chapter is the municipal organisation of Malmö, Sweden. While not representing a traditional "high-risk industry" commonly studied in safety research, it is important to point out that many public sector organisations are used to manage complex and surprising events as part of their daily work. For many municipal workers, especially those in departments responsible for critical activities such as health care and child care, it seems that the ability of "managing the unexpected" is so ingrained in the organisation that they seldom reflect upon it. This is not to say, however, that an effort of analysing and nurturing resilient performance is unwanted or unneeded, but rather reflects the fact that much of the abilities to perform resiliently in complex situations builds on tacit knowledge, which underlines the importance of not marginalizing such local know-how.

In the autumn of 2015, an event commonly referred to as "the refugee crisis" unfolded in Sweden (however, it is important to recognise that, although this phrase has gained traction in Sweden, the word "crisis" in this context is far from politically neutral). During the year, 163,000 people applied for asylum in Sweden, of which approximately 35,000 were unaccompanied minors. In the short period between October and November, the number of asylum seekers amounted to almost 80,000. The municipality of Malmö in southern Sweden was one of the municipalities receiving the largest number of refugees, primarily due to its geographical proximity to neighbouring countries.

For the municipality, the event in itself was unforeseen (although signals of its imminent occurrence could have been detected – but as always – this is easier to conclude with hindsight). During the municipality's acute handling of the situation, shortcuts were taken, corners were rounded, and from time to time delivery of support and services to people was far from ideal. Yet, given the circumstances, the municipal organisation showed a remarkable ability of flexible and adaptative performance under severe pressure in the way it managed to cater for the most immediate needs of the arriving refugees.

From the perspective of RE, it is tempting to ask whether the municipal organisation of Malmö performed resiliently during this event? This is obviously an intriguing question to answer. Degerman et al. (2018) have tried to address this question by studying to what extent Malmö municipality, and some of the other organisations involved, showed instances of adaptation to evolving circumstances and needs during the event. One of their findings pointed at a discrepancy between the actual challenges facing operational staff and higher management's understanding of this practical work. While operational staff managed the situation by adjusting or abandoning some of the existing routines, there was a belief among upper management that the situation could be managed through ordinary work practices. Based on these findings, the authors recommended increased knowledge and understanding at the management level of the constraints and conditions facing frontline staff, motivated by a need to ensure that management do not undermine the operational work.

We agree on the urgency of such efforts and the need for management to better understand operational work. When our research in the municipality of Malmö was initiated less than one year after this event, our scope was not restricted to the abilities of managing the situation of a sudden increase in the influx of refugees, but more broadly about increasing the ability of assessing the organisational capacities of anticipating and managing any type of unexpected event or crisis. Yet, we did have the challenge in mind of trying to address the divide between conditions facing operational staff and management understanding of the situation. As part of this, one of our aims was to work towards an alignment of "work as done" by operational staff and "work as imagined" by management in the municipal departments. In addition, we believe that increased knowledge among operational staff regarding "work as done" performed by other units can also increase organisational capacities to perform resiliently; especially in systems where there are large dependencies between different operational units. This is often the case within and between modern organisations performing vital social functions, for example, due to organisational reforms aiming at increased efficiency, which will be described in more detail below. While the municipality of Malmö was used as our case, we believe this is a more general challenge facing large organisations performing vital societal functions, and not only in the public sector.

2 Structural Secrecy

RE places significant attention on the ability of "knowing what goes on", commonly expressed as the ability to "monitor" internal as well as external conditions and processes. Contrasting with our experiences from Malmö, we can conclude that it appears that this ability is at odds with the way most (public sector) organisations are designed. Rather than providing opportunities for people within organisations to actually know what is going on through active channels of communication and information sharing, the *modus operandi* of most organisations is to work towards increased specialisation and professionalisation. While these mechanisms provide significant value to organisations, such as increased efficiency under normal operations and creation of more manageable units of work, they simultaneously lead to challenges related to managing complexity. In situations where more effort is devoted to dividing areas of responsibility between multiple actors and departments than to coordinating the different units, problems may emerge when responsibilities fall between the cracks (Heath & Staudenmayer, 2000). Moreover, when the goals or time perspectives of different units or organisations are not aligned, there is a risk of individual actors working in ways that are collectively detrimental, or even at cross purpose (e.g. Ostrom, 1999; Woods & Branlat, 2011). In addition, specialization and efforts devoted at increased efficiency also depletes an organisation from overlaps, slack and redundancies, all of which are vital for managing events that stretches the organisational capacities. Operational dependencies between organisations and organisational units also create conditions for cascading effects to arise throughout the nested operations.

When efficient means of communication across departments are not in place, the organisation suffers from a form of structural secrecy in terms of hindering information from spreading between those who have a need to share it. Vaughan (1996: p. 250) has elaborated on the concept of structural secrecy as a systematic undermining of attempts to know and interpret situations in organisations:

> Secrecy is built into the very structure of organizations. As organizations grow large, actions are, for the most part, not observable. The division of labor between subunits, hierarchy, and geographic dispersion segregate knowledge about tasks and goals. Distance – both physical and social – interferes with the efforts of those at the top to 'know' the behavior of others in the organization – and vice versa. Specialized knowledge further inhibits knowing. People in one department or division lack the expertise to understand the work in another or, for that matter, the work of other specialists in their own unit.

Structural secrecy is especially challenging in cross-organisational settings, or in organisations with multiple and diverse areas of responsibility with limited amount of communication between the different parts, such as in a municipal organisation. The municipality of Malmö employs about 25,000 people and the areas of responsibility range from child care, education, cultural activities, water and sewage, elderly care, and urban planning. Given the size and complexity of this organization, "knowing what is going on" practically becomes impossible in any meaningful sense. Each area of responsibility is effectively managed in separate silos with limited interaction across departmental borders. Yet, a serious stressor affecting one department calls for swift response that often requires action by several municipal departments as well as by external actors. For example, several departments may have dependencies to some common support system that, if affected by a stressor, may give rise to serious impacts on several departments. As such, the classical dilemma of being able to shift between centralization and decentralization is prominent for managing unexpected events.

The complexity arising from structural secrecy is not only seen in the organisation of day-to-day work. It is also seen in efforts to understand organisational resilience, risk, safety, vulnerability, continuity, etc. In the case of municipalities in Sweden, a large range of analytical activities are performed to understand organisational resilience and risk. Numerous perspectives, such as accidents, crises, workplace conditions, climate risk, financial risk, cyber risk and information security, are in many cases addressed by loosely coupled activities. Coordinating these overlapping analytical efforts in a way that makes sense for the frontline operational staff, therefore, constitutes a complex task that further exacerbates the issue of "knowing what goes on".

3 Other Forms of Secrecy

Structural secrecy is merely one of the mechanisms that gives rise to secrecy in complex organisations. In the case of the municipality of Malmö, we have observed at least two other mechanisms of secrecy, here labelled "intentional secrecy" and "semantic secrecy". These two mechanisms are quite different from structural secrecy but give rise to similar effects in terms of increased difficulties of knowing what is going on.

Intentional secrecy arises from an aversion of organisations or organisational units to share information about how they function (what they do, what they depend on, what resources they possess, what their capabilities and vulnerabilities are, etc.). Such an aversion may have multiple reasons, and Månsson (2019) has provided an overview of factors that inhibit sharing of risk information. The most obvious example is the risk that a malevolent actor may use information about the functioning of an organisational unit in order to induce harm. With increased attention in Sweden devoted to sabotage and scenarios of heightened state of alert, many public organisations have become much more aware, and restrictive, when it comes to what, how, and with whom information is shared, hence, reducing the potential of increasing knowledge about "what goes on". In addition, a person may be less inclined to share information about the limits to an organisation's resilient performance due to a fear that the organisation will blame the person for spreading information that will make the organisation look bad. This can be seen as somewhat equivalent to people in safety-critical organisations unwilling to share information about incidents due to a fear of being punished for not complying with formal procedures. Although there is no universal remedy to the issue of intentional secrecy, an obvious solution is to increase the level of trust between organisational units and organisations, which can be addressed by efforts to ensure information is treated in a secure and sensible way.

Semantic secrecy arises from efforts of actually trying to share information about risk and resilience. However, this can be challenging in a multi-actor setting. For example, if aspects or indicators of risk and resilient performance is described in incompatible ways, organisational units may have difficulties understanding this information and integrating it with other information. This phenomenon has been investigated at length in Månsson et al. (2015) and has been labelled uncommon categorisation by Kramer (2005). Increased standardisation is one way forward to reduce the negative effects of semantic secrecy; however, standardisation or harmonisation can be difficult to implement in a multi-actor setting, and it has several drawbacks when it comes to its effects on managing risk.

4 Increasing Knowledge About "What Goes on"

In a municipal setting, the availability of formal or informal channels for interaction between departments may be decisive for the outcome of an emerging event. For example, in the case of flooding or fire in an elderly care or child care facility, knowledge about what redundant buildings are available in other departments for temporary evacuation is highly valuable. Helping an organisation to develop cross-departmental knowledge that can be utilised when unexpected events strike may, therefore, be crucial. This calls for efforts aiming at transforming individual workers' tacit knowledge about connections between organisational units to more explicit knowledge accessible to a larger part of the organisation.

While the task of developing such knowledge base may fall on the responsibility of a safety professional, this is not something this person can do on his/her own. To obtain such knowledge, expertise among frontline workers need to be exploited, as these people's expertise on what actually goes on in the organisational units is highly valuable as a means to understand how more resilient performance can be nurtured. To gain such rich and local knowledge in the municipality of Malmö, a decentralised approach was used where a method for collecting data about each municipal department's activities was developed and implemented (further described in Hassel & Cedergren, 2017).

While the compilation of such data is highly valuable, our experience shows that this is not accomplished in practice without significant challenges. Firstly, while formal efforts aiming at building a more resilient organisation may make sense to most staff, this is outside most employees' main work tasks, which means that time and commitment is limited. These constraints are even more prominent for the management level in the municipal departments. Secondly, the procedure for compiling information necessary to gain an overview of the organisation's critical activities and the connection between different organisational units should not require a high level of knowledge or training, or make use of academic terms and concepts that have no intuitive meaning to the staff involved. Even what appears to be small hurdles, such as problems for frontline staff to understand what information that needs to be collected, may create significant frustration that threatens to bring the whole process to a halt. Thirdly, in order to have an actual value to management, in broader terms than only for the staff involved in the data collection process, the outcome from such process should be possible to present in an attractive format that gives immediate insights about the organisation's work practices. In our view, a graphical representation is more suitable than written reports for this purpose. Moreover, since information about the activities of municipal departments may become rather comprehensive and complex, some kind of interactive format is also preferred.

In our work in Malmö, we developed a method aimed at providing awareness among organisational members about capacities and vulnerabilities of each department and their dependencies to other departments as well as to external actors and resources. The method was based on principles from the area of continuity management (see ISO 22301), and method development was conducted in close collaboration with end users in the municipal departments. A reference group consisting of staff from 7 out of 14 municipal departments was created as a means to collect input throughout the method development process, which was carried out by iteratively testing and evaluating one step of the method at the time. Based on the feedback provided by the end users in the reference group, adjustments were made until a final method was developed that was simple enough for end users to use with minimal external support, while still sufficiently detailed to give rich data about each municipal department. The method development took place over a period of approximately 2 years through a series of workshops, which was followed by an implementation phase in the departments.

An important part of the developed method was about mapping functional dependencies between departments, which provides a value in a very large organisation where people struggle to get an overview of all parts of the organisation (including their own department). Assessing activities that are time-critical, that is, that may give rise to unwanted consequences if they cannot be undertaken, represents a way of transforming tacit knowledge among those who work with these activities to more concrete knowledge accessible to larger parts of the organisation. The process takes a number of steps, as follows:

1. *Identifying what the organisational unit actually does (and not what it is thought to do).*

   This first step aims at arriving at a description of the various activities going on in the organisational units, based on local knowledge from those working in the units. Such description needs to be of a feasible level of detail; sufficiently detailed to describe what actually goes on, but not more detailed than a manageable number of activities are identified, given available time and resources for this task.

2. *Assessing which of these activities that quickly may lead to negative consequences.*

   While all activities in an organisation are conducted for a reason, and as such, are necessary in some way, they are not equally time-critical. This step, therefore, aims at investigating how time-critical the various activities are, and in doing so, sorting out activities that have significant amount of slack. This is done, for one activity at the time, by assessing how long it would take until the activity – if it cannot be undertaken – gives rise to serious consequences. In order to capture expected as well as unexpected events, focus of this process is not on what may cause these disruptions, but only how quickly an impact on the activity may create serious impacts on the organisation. At first, such estimations are tentative, since the knowledge about how other actors depend on the particular activity may be limited. Then, as more knowledge about downstream consequences of disruptions are gained from information shared by other actors, these estimations should be re-evaluated.
3. *Identifying what the activities are dependent on.*

   The undertaking of each activity depends on a number of resources, such as personnel, IT systems, and facilities. The most important resources are described in this step, and the strength of these dependencies are assessed, for example, in terms of how quickly and to what extent the activity is affected if the resource is not available. In order to reduce the semantic secrecy, a common categorisation of dependencies are used across the municipal departments.
4. *Identifying what backup solution that exists.*

   Most resources that the organisational activities are dependent on typically have some kind of backup solution. For example, in situations of staff shortage, a pool of extra staff can be used, or if an automated procedure is out of order, manual routines can be used. In this step, such backup solutions are described, and the robustness of these solutions are assessed, as well as which actor is responsible for the backup solution.
5. *Illustrating the results in a holistic way that can be communicated to the management level as well as the organisational units.*

   While the analysis conducted in the previous steps is important in itself, it is equally important to present the results in a way that helps to create overview and understanding of the capability of the organisational units. Such an overview can then be communicated to the management level as well as to each organisational unit in order to create an understanding of how activities in one department is connected to activities in other departments through various dependencies. Figure 1 shows how this may be done based on the data collected in Steps 1–4. Activities belonging to each municipal department, including its time-criticality, is depicted along with their dependencies (including their criticalities). These steps are conducted in each department and compiled in the form of a network for the entire municipal organisation as shown in Fig. 2. As such network grows in size, it becomes necessary to highlight some aspects of the whole network, and understanding is facilitated by possibilities to interactively explore and analyse the network.

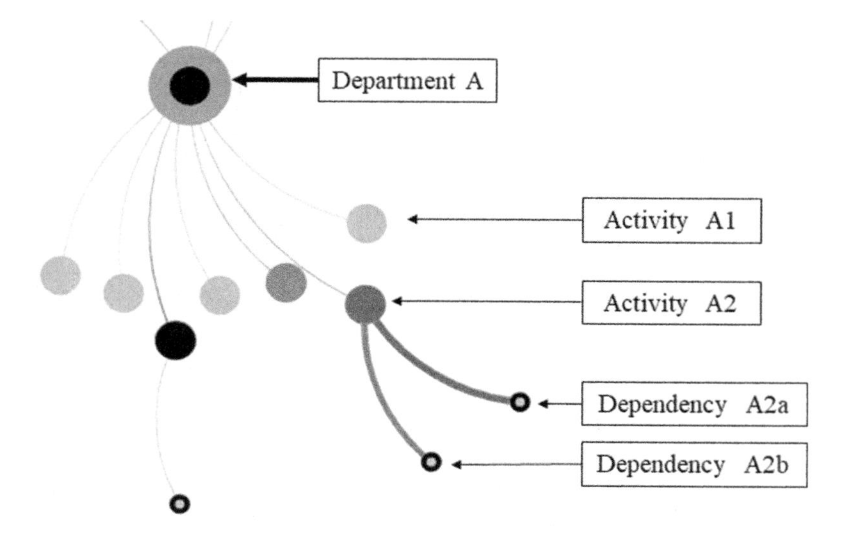

Fig. 1 Illustration of Department A's activities and dependencies

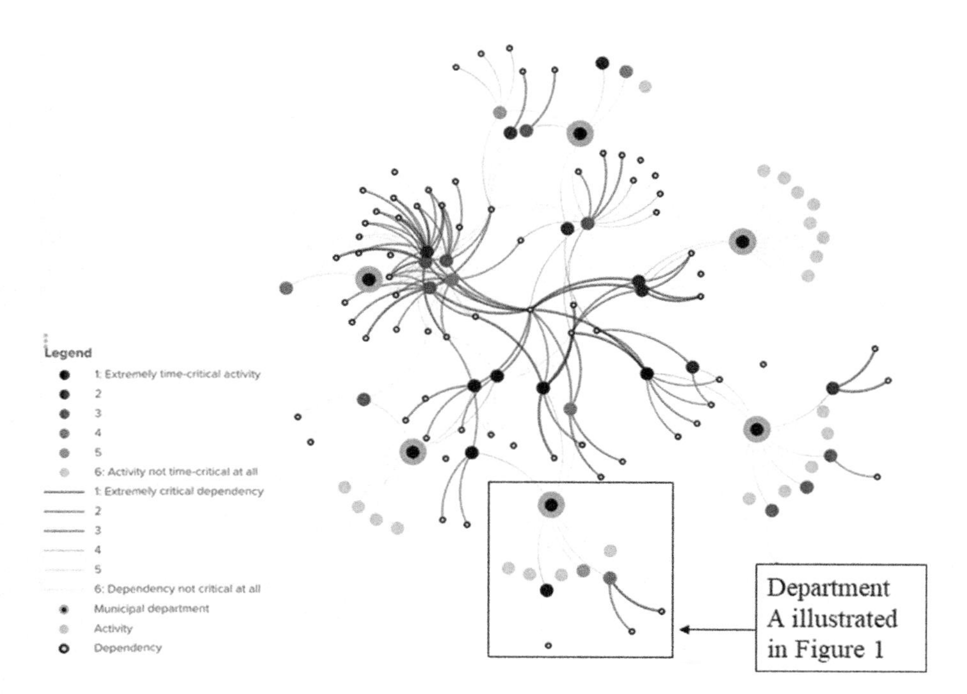

Fig. 2 Illustration of several municipal departments, their activities and dependencies, including those dependencies that several departments have in common

In Fig. 1, some of the data generated from using the method in Department A is illustrated. As can be seen in this figure, Department A is responsible for a number of activities, two of which are referred to as Activity A1 and A2. Due to confidentiality, we cannot provide information about what these specific activities are, but for the purpose of illustration, it can be assumed that this department is responsible for road maintenance and other related maintenance services. Activities A1 and A2 could then correspond to "street cleaning" and "snow clearance", respectively. The darker colour of Activity A2 indicates that this activity is more time-critical compared to A1 (i.e. if the activity cannot be undertaken, it will more quickly lead to negative consequences as defined in a number of pre-established consequence scales). Moreover, Fig. 1 shows that for Activity A2, the end users have specified two dependencies, which could be "Vehicles for snow clearance" (A2a) and "Fuel" (A2b), respectively. While not illustrated in Fig. 1, the method also allows for specification of backup solutions, which for dependency A2b could be "Fuel depots" and "Contracts with several fuel suppliers". Figure 2 shows data generated by several departments. As can be seen in this figure, dependencies that are common to several departments can be identified when the results are aggregated in the form of a network.

While our work is only briefly outlined in this chapter, some important insights can be highlighted. Firstly, from the kind of representation illustrated in Fig. 2, increased knowledge and understanding can be gained about which organisational tasks and activities are more time-critical, and which ones are less so (darker dots represent activities that are more time-critical). This gives an indication of where to boost the organisation's abilities of anticipation and response to potential stressors (rather than seeing these abilities as something that are required all across the organisation, which is not a realistic approach in most organisations due to limited resources). Even though we share the view of adaptive capacities among frontline staff as an invaluable resource to manage events under time pressure and economic constraints, we believe that these abilities can be strengthened by demonstrating to management level where they are most acutely needed; and also, by demonstrating to each organisational unit how they depend on, and affect, other organisational units. Secondly, Fig. 2 provides insights about the criticality of resources that the various activities are dependent on, and the strength of the backup solutions available (the thicker and darker link between an activity and a resource means a more critical dependency). This constitutes vital information about whether some resources are critical for a range of activities as well as whether some activities may be prone to disruptions due to many critical dependencies. Thirdly, Fig. 2 shows what resources that many organisational units have in common. This means that one unit may rely on a specific resource that another unit also relies on. In situations where this resource is not available, both units are negatively affected. Moreover, different organisational units (that may not be dependent on each other on a daily basis) may make changes or remove slack and redundancies that may have become obsolete from their changing needs or responsibilities, but that others may count on. While this knowledge might exist among the members of an organisation, it is rarely compiled and aggregated on an overall level, meaning that the main value of this kind of approach is to break up the patterns of secrecy to connect tacit knowledge that is spread throughout the organisational units. Making such information more readily available in the organisation may also increase the possibility to make accurate judgements about how critical different activities and resources are for the organisation as a whole. In addition, it can spark the need for dialogue between organisational units when it comes to organisational changes that may have repercussions on other units or actors.

5 Conclusion

Continued efforts of taking theoretical aspirations to practical application constitutes a remaining challenge for RE. In this chapter, we have briefly shared some output from a collaboration with the municipality of Malmö for a period of more than 3 years that has provided insights into the needs and constraints to become better equipped for managing both daily tasks as well as disruptions and surprises. Our starting point for this chapter was a desire to address problems related to secrecy in large (public sector) organisations as a way of strengthening the ability of resilient performance once disruptions or stresses occur. To deal with this problem, it is essential to find ways of aligning diverging views on work as done versus work as imagined (particularly between management and frontline staff), combined with ways of sharing insights about how critical these activities are, and how they are interconnected. In our view, this gives us a chance to take necessary proactive actions and make deliberate investments in supporting the organisation's readiness to adapt in a more thoughtful way than solely relying on the ability of heroic frontline operators to make the necessary adaptations and trade-offs in unexpected events as they occur. As such, this type of effort lays the groundwork for nurturing an ability to perform resiliently in suddenly emerging situations that are outside the organisation's normal operations.

References

Alexander, D. E. (2013). Resilience and disaster risk reduction: An etymological journey. *Natural Hazards and Earth System Sciences, 13*(11), 2707–2716.

Birkland, T. A., & Waterman, S. (2009). The politics and policy challenges of disaster resilience. In C. P. Nemeth, E. Hollnagel, & S. W. A. Dekker (Eds.), *Preparation and restoration. Resilience engineering perspectives* (Vol. 2, pp. 15–38). Ashgate.

Boin, A., Comfort, L. K., & Demchak, C. C. (2010). The rise of resilience. In A. Boin, L. K. Comfort, & C. C. Demchak (Eds.), *Designing resilience: Preparing for extreme events* (pp. 1–12). University of Pittsburgh Press.

de Bruijne, M., Boin, A., & van Eeten, M. (2010). Resilience: Exploring the concept and its meanings. In A. Boin, L. K. Comfort, & C. C. Demchak (Eds.), *Designing resilience: Preparing for extreme events* (pp. 13–32). University of Pittsburgh Press.

Degerman, H., Bram, S., & Eriksson, K. (2018). Resilient performance in response to the 2015 refugee influx in the Øresund region. In S. Haugen, A. Barros, C. van Gulijk, T. Kongsvik, & J. E. Vinnem (Eds.), *Safety and reliability – Safe societies in a changing world. Proceedings of ESREL 2018, Trondheim, Norway* (pp. 1313–1318). CRC Press.
Haavik, T. K., Antonsen, S., Rosness, R., & Hale, A. (2019). HRO and RE: A pragmatic perspective. *Safety Science, 117*, 479–489.
Hassel, H., & Cedergren, A. (2017). A method for combined risk and continuity management in a municipal context. *Proceedings of the 2017* European Safety and Reliability Conference ESREL, Portorož, Slovenia.
Heath, C., & Staudenmayer, N. (2000). Coordination neglect: How lay theories of organizing complicate coordination in organizations. *Research in Organizational Behaviour, 22*, 153–191.
Hollnagel, E., & Woods, D. D. (2006). Epilogue: Resilience engineering precepts. In E. Hollnagel, D. D. Woods, & N. Leveson (Eds.), *Resilience engineering: Concepts and precepts*. Ashgate.
Hollnagel, E. (2006). Resilience–The challenge of the unstable. In E. Hollnagel, D. D. Woods, & N. Leveson (Eds.), *Resilience engineering: Concepts and precepts*. Ashgate.
Hollnagel, E. (2008). Preface: Resilience engineering in a nutshell. In E. Hollnagel, C. P. Nemeth, & S. W. A. Dekker (Eds.), *Remaining sensitive to the possibility of failure. Resilience engineering perspectives* (Vol. 1, pp. xi–xiv). Ashgate.
Hollnagel, E. (2009). The four cornerstones of resilience engineering. In C. P. Nemeth, E. Hollnagel, & S. W. A. Dekker (Eds.), *Preparation and restoration. Resilience engineering perspectives* (Vol. 2, pp. 117–134). Ashgate.
Hopkins, A. (2014). Issues in safety science. *Safety Science,* 67, 6–14.
International Organization for Standardization. (2012). *ISO 22301 societal security: Business continuity management systems: Requirements*. Geneva, Switzerland.
Kramer, R. M. (2005). A failure to communicate: 9/11 and the tragedy of the informational commons. *International Public Management Journal, 8*(3), 397–416.
Månsson, P. (2019). Uncommon sense: A review of challenges and opportunities for aggregating disaster risk information. *International Journal of Disaster Risk Reduction, 101149.*
Månsson, P., Abrahamsson, M., Hassel, H., & Tehler, H. (2015). On common terms with shared risks–Studying the communication of risk between local, regional and national authorities in Sweden. *International Journal of Disaster Risk Reduction, 13*, 441–453.
Ostrom, E. (1999). Coping with tragedies of the commons. *Annual Review of Political Science, 2*(1), 493–535.
Pendall, R., Foster, K. A., & Cowell, M. (2010). Resilience and regions: Building understanding of the metaphor. *Cambridge Journal of Regions Economy and Society, 3*(1), 71–84.
Vaughan, D. (1996). *The challenger launch decision: Risky technology, culture, and deviance at NASA*. The University of Chicago Press.
Woods, D. D. (2015). Four concepts for resilience and the implications for the future of resilience engineering. *Reliability Engineering and System Safety, 141*, 5–9.
Woods, D. D., & Branlat, M. (2011). Basic patterns in how adaptive systems fail. In J. Wreathall (Ed.), *Resilience Engineering in practice: A guidebook* (pp. 127–144). Ashgate.
Woods, D. D., & Hollnagel, E. (2006). Prologue: Resilience engineering concepts. In E. Hollnagel, D. D. Woods, & N. Leveson (Eds.), *Resilience Engineering: Concepts and precepts* (pp. 315–325). Ashgate.
Woods, D.D. (2003, October 29). *Creating foresight: How resilience engineering can transform NASA's approach to risky decision making*. Testimony on the future of NASA for Committee on Commerce, Science and Transportation. John McCain, Chair, Washington D.C.

The Second Step: Surprise Is Inevitable. Now What?

Beth Lay and Asher Balkin

Contents

The ontological underpinnings of resilience engineering focus on the experience of surprise as an inescapable component of existence in the complex adaptive universe (Hollnagel et al. 2006). The resource limitations of all systems (especially cognitive systems) force the units of those systems (agents) to create and rely on models of the world. These models are necessary and necessarily incomplete – they are simplifications of the world on which they are based. Because they are simplifications, there will be inaccuracies in their predictive power. This will lead the agents – users of the models – to experience surprise (Hollnagel et al. 2006). Although past attempts to produce thorough and exhaustive lists of all potential outcomes (robustness exercises) have produced significant results in controlled settings or where large recourse endowments are available, the benefits of those techniques are architecturally resource-limited by the mitigative mechanisms which field the creativity of the practitioners devising the scenarios. Such attempts to "outthink" the universe by defining each potential specific instance are necessarily limited in their ultimate

B. Lay (✉)
Lewis Tree Service, Henrietta, NY, USA

A. Balkin
The Ohio State University, Columbus, OH, USA

C. P. Nemeth, E. Hollnagel (eds.), *Advancing Resilient Performance*,
https://doi.org/10.1007/978-3-030-74689-6_11

operational effectiveness. Consequently, capacities to deal with the inevitable surprises must be developed and maintained when and where the work is done.

Tales of nearly catastrophic events are typical of utility-line clearance work. "I was felling a large, dead, locust tree surrounded by slippery, moss-covered ground. The tree started to fall in the intended direction then the top broke out and was coming right for me. I fell as I scrambled to get out of the way. I barely escaped on my hands and knees…" recounted a 30-year veteran manager of tree work.

The highly variable nature of clearing trees from powerlines, major differences in work site geography, environmental conditions, arboreal architecture, etc. with few available options to directly control those inconsistencies, results in overall variability of the work being exceedingly high. When combined with high rates of worker turnover, which make the training and retention of skilled workers difficult, line clearance routinely ranks as one of the most dangerous jobs in the United States. The most common and serious risk faced by the line clearance industry is being struck by a tree or limb, often resulting in serious injury or fatality. Informal polls of tree workers reveal that most have only narrowly escaped being hit by falling branches. Despite the persistence of this condition, industry training requirements have remained unchanged for decades (1910 USC 226a).

Typical responses have been, and continue to be, advice to "be careful" or policies to "stay out of drop zone (the area under the tree where falling limbs are most likely to land)".

As this chapter began, we introduced the idea of resource limitation. Because cognitive agents are limited in means and mechanisms to address any task or meet any challenge, they are forced to work with models – simplifications and incomplete conceptual constructions of the world around them. The need to build and use simplifications makes cognitive agents susceptible to model surprise. Model surprise, the most common type of surprise experience, occurs when the mental model of a cognitive agent fails to accurately predict an upcoming event or condition. The complexity of the world in which line-clearance workers operate, and the rate of speed at which the world changes (transition from a stable tree to an unstable/falling branch), as well as foundational characteristics of the work tasks which often involve work wherein large parts of the work task (inside of the tree) are mostly or totally opaque to the worker, combine to encourage the conditions of frequent model surprise.

Surprise is a normal part of work, and surprise is a ubiquitous part of highly variable work.

1 Utility Line Clearance as a Natural Laboratory

Principally, line-clearance tree work involves pruning, trimming, and removing hazardous trees along the routes of electric powerlines to improve power system safety and reliability by diminishing or removing the hazard presented by tree limbs falling on or in other ways contacting the lines and thereby damaging equipment and/

or causing outages (service interruptions to utility customers). The confluence of multiple factors makes this work domain a near-ideal natural laboratory to the exploration of systemic and organizational resilience. Most obviously, the occupation is definitionally high risk and high consequence. The ever-present risk of falling tree limbs combined with the danger of high voltage line contact has made tree work one of the most injury- and fatality-prone jobs in North America for many years running (ISHN, 2020). Furthermore, several specific factors can be identified which exacerbate the need for resilience in practice as line-clearance tree work is performed.

High degrees of worksite opacity – Unlike engineered systems, which are often (or at least hopefully) designed with the obligations of human maintenance in mind, natural systems perform their own maintenance behaviors, the operation of which is often not obvious to human operators nor are such systems seemingly designed for the reequipment of human maintenance functions. Engineered systems, factories, computer networks, etc., are, at least in theory, constructed in such a way as to permit regular system maintenance. In this context, the trimming of trees and hazard-tree removal should be considered a normal power system maintenance function. As a result, the site of tree work (the grounds) as well as the subject of the work, the trees themselves, often conceal hazards which, in engineered systems would be made more obvious. Unstable soil, insect and animal nests, and internal tree decay are just a few examples. Wind can blow limbs into the lines triggering an outage.

Little control over worksite conditions – Related, but operationally different from the concept of work subject opacity, is the level of control the worker team has over the overarching conditions of the worksite. These would include components such as weather as outdoor labor cannot be controlled, but also factors relating to the geography of the area such as steep slopes or the presence of homeowner vehicles/homes or other more specific architectural features such as backyard swimming pools which may make accessing overhead powerlines more difficult. Line clearance companies typically rely on worker judgement of when to stop work since wind speed changes quickly and direction of wind matters. Adding to this is that when workers "call the day" due to wind or rain, they often do not get compensated; incentivizing continuing work in less than optimal conditions.

More extreme instantiations of this include the inviolability of the powerlines and poles themselves. While the safety of the trimmer operation would be greatly improved if crew were able, as they may be in a more confined industrial setting, to deactivate/deenergize the transmission/distribution cables, or better yet, temporarily lower them to the ground, such actions are often not practical in this work setting. Additionally, unlike construction sites, or even road-repair workers, line-clearance crews often operate with little ability to exclude the public from the work area for any significant length of time – often adding to the time pressure to complete the task. While not an exhaustive list, these few examples serve to demonstrate the minimal degree of control the work teams have over their worksite, particularly as it might be compared to mare traditional industrial/manufacturing settings.

Rapid redeployment/worksite relocation – Utility line clearance is fundamentally a mobile work environment; success in completing a day's work likely means moving to multiple new sites, with potently new hazards, each day. Contrasted with work domains where the site of work is fixed, even for several days, if not weeks or months as is often the case in construction and development projects, line clearance crews have little time to inspect, consider, and adjust to new worksites and the hazards they hold.

High levels of workforce turnover – Compounding the issues above, are high levels of worker turnover which makes the development of experienced workers difficult. Consequently, work teams frequently have new members which, in addition to operating with workers of reduced technical and operational skill, has been shown by internal data and accident analysis to increase the likelihood of accidents, and decrease inter-predictability among team members – a critical contributor to work-team resilient performance.

2 Work as Done: The Techniques of Resilient Performance in Line Clearance Tree Crews

While theoretical approaches to the exploration of resilience are useful for discussing, describing, and exploring system behavior, the most significant insights which might be gained from observing and working with line clearance tree work are practical strategies.

The Story of the Elephant – Perspective Shift: Although it has long since spread beyond its original Buddhist and Hindu heritage, the story of the blind men and the elephant presents a striking illustration of the notion that all perspectives are both revealing and obscuring. The most abridged version of the story has several blind or blindfolded men standing next to an elephant touching different parts of the same animal. The man touching its leg believes he holds onto a tree trunk; the man holding the ear thinks he is touching a fan, the man holding the elephant's tail claimed to be grabbing a rope. While many versions of the tale have developed and been told since its believed inception in the first millennium BCE, all reveal the same truth: the perspective of each person was limited by the information they held. Furthermore, because of the highly physical nature of the experience, what each person believed to be true of about their situation was highly depending on where they were in the scene, and thus what part of the animal they had access to (Saxe, 1885).

Seeking to experience the benefits of a perspective shift, the craft work crews, their supervisors, and safety advisors, have devised several methods which encourage the shifting of perspective, the first of which is the most direct: a physical change of location for the observer. During the pre-work hazard assessment and work planning process, recent changes in the work planning process have encouraged craft crew leaders to recognize explicitly what the work crews have learned experientially for some time – the work never looks the same from the bottom of the trees as it does from the top.

Though physical movement has been undoubtably helpful to the crews, strategies to shift *cognitive perspective* have been implemented and shown results as well. One of the most productive of those mechanisms has been a program to encourage a general foreman to request a second opinion on difficult work situations. Most relevant is the decision to invite an additional general foreman into the situation is the recognition that the decision is not made to add additional direct work resources to the team. Rather, the introduction of the additional general foreman adds *cognitive resources* to the team. While the introduction of the new agent does add a valuable experiential capacity to the team, just as, if not more important, is the forced change in perspective that the new foreman will bring. As a result of not having been involved in the initial planning process, the new foreman is likely to recognize new elements of the work situation that may have been overlooked by those already onsite.

Respond to emergent risk and uncertainty Practices to manage emergent risk are critical for highly variable work. Effective practices include mechanisms for pausing work/stop-work authority (for Lewis Tree of West Henrietta, NY, USA, it is called "Press Pause" and "All Stop"), followed by a process to make sense of the situation then replan. Organizations often hone in on decision-making when considering emergent risk. We offer that this focus may be misplaced/inadequate.

Let us start with stopping work. We tend to consider calling an All Stop to be a simple, worker decision. We give all the Authority to Stop Work. We ask workers to "stop when unsafe". After an incident, we ask: "Why didn't they stop work? It's clear there was an unsafe situation!"[1].

The problem is that this is based on several unrealistic assumptions (Weber et al., 2018):

1. A situation is clearly safe or unsafe.
2. Warning signs are always present and easily visible.
3. Stopping co-workers is always possible.
4. The task process is stable at every step; therefore, it is always safe to stop.

Leadership has a significant influence on whether people stop work. Factors that prevent or support stopping are primarily social as compared to individual behavioral choices. Leaders can act to stop work in a number of ways:

- Invite people to call an All Stop. Especially, the team lead just before performing a task "If you see something, call an All Stop".
- Ensure that asking for help and bringing in a second opinion is seen as a strength.
- Explore worker's views on safety, risk, danger, and tasks to continue versus stopping.
- Make it clear as possible in which situations you want people to stop work and escalate. Be specific in defining when you want people to call an All Stop: limb stuck aloft, leader/trunk did not break loose as expected.
- Acknowledge there may be ambiguity "If you are uncertain, if something does not seem right – it's ok to call an All Stop." Waiting until you are certain may be too late.

- Always say "thank you". Never question if calling the All Stop was the right thing to do.
- The "Press Pause" tool includes practices to make sense of the situation. Questions to probe uncertainty and surface risks: what's different that adds risk? etc.

We return to Hollnagel's four potentials of Resilience Engineering (anticipate, monitor, respond and learn) to explore how we can respond resiliently in the face of surprise considering that:

- Surprise is fundamental to the human experience in life.
- Surprise is fundamental to the human experience in work.
- Surprise, as an experience, is unavoidable; however, surprise, as an experience is manageable.
- Surprise can be both positive and negative.

Characteristics of surprise include rate, kind of surprise, and detectability. We asked "how were you surprised?" after people were struck or almost struck by a limb or a tree; their answers help us understand the characteristics of surprise. We learned that it happens fast (tree or limb falls fast), it "goes bad quick" (little warning before failure). The path of the falling limb or tree becomes less predictable when the limb or tree is curved or oddly shaped (nonlinear mass distribution). The limb or tree is rotated (spun or twisted), bounced or ricocheted (off other branches, an adjacent tree, or the ground), or fell in an unplanned direction. Detectability was low in many cases such as when there was decay inside the tree or cracks or other flaws high in the tree.

3 Anticipate, to Prepare for Surprise

Anticipation is a forecasting behavior which attempts to avoid the need to respond to surprise by forestalling the surprise itself. Superficial interpretations of the relationship between anticipation and surprise might suggest surprise to be consequence of incomplete anticipation. While the argument retains a logical truth, it lacks any utility as our anticipatory models of the world are, even under the best of circumstances, inaccurate. Or more directly, they are necessary and necessarily incomplete. Their necessity lies in their existence as a pre-condition to action while their incompleteness derives from our perspective-driven limitation of access to information and our cognitive limitation in processing and sensemaking (Lynam & Fletcher, 2015 and Weick et al., 2005).

We can expand our experiences of the world through the stories of others. Story-style case studies set the scene including clues that things might take a turn for the worse. Discussions with frontline workers often yield insights such as: "There was a sense of urgency to get this job done. The lead foreman was hired about 2 weeks earlier and only had a handful of working days under him. This was a fairly difficult to tree to remove. His new plan was to take the top 60 feet of this oak tree out with one cut." Progressive questions such as "What were the warning signs?" "What do you think happened next?" help us practice anticipation.

When we include the strong emotions that go with being surprised in the story, we will remember the event almost as if it had happened to us.

> "I moved control lever up but the bucket started going down, toward the power line." I thought "I could die here".

Telling a story firsthand is powerful in triggering emotions. This story was relayed by a 21-year-old safety team member, Moises, who was a groundman when the event happened, on an Operations Leadership safety call:

> Moises' story: the trimmer realized I was underneath him, he yelled "watch out!" I froze and started to look up but before I could raise my head all the way, the limb hit the front brim of my hard hat. .... I was completely shocked. The limb was 13" in diameter.

This story triggered a visceral reaction among the listeners, bringing forth how close this was – within inches – of losing Moises. The story also revealed the reality that most likely there are many close calls with being struck by limbs or trees that are never reported. Personal stories like this help create the space for people to fill in the details with their own imaginations; they expand mental models. Case studies are short and delivered by a general foreman or another team member, during morning meetings, standing in a parking lot. A "storytelling coach" (local newspaper editor) collaborated to design a process that includes three facets:

- Mine stories, including close calls: probe the extremes of experience (worst, best, first, last, hardest, proudest, scariest). Get to the emotions
- Craft stories: lead quickly with something that matters, engage the senses, build suspense
- Tell stories: prepare ahead of time, prime audience to listen for clues, share as a conversation

Leadership talks a lot about creating safety and learning from work that goes well. Workers began sharing good catches; stories of anticipating what could go wrong and making adjustments in advance.

> We anticipated the rigging point could fail so we added an extra rigging point. The 1st rigging point did fail because the tree was decayed...

> I asked another general foreman to come take a look. We noticed the neutral wire had only a couple of strands holding it together.

4 Ambiguity as an Indicator

Surprise is often preceded by a feeling of uncertainty, which may be triggered by ambiguity. Ambiguity may take the shape of a lack of clarity regarding the present conditions (lack of confidence in one's own model/understanding of the situation) or closely related, uncertainty regarding the most efficacious next steps. While the mere presence of ambiguity ought to be considered a signal, and often is by experienced reflective practitioners, more often signals of ambiguity are de-emphasized, diminished, disregarded.

Nevertheless, engaging the unknown by acting to learn is critical for making better decisions in uncertain situations. Waiting until things are certain is often too late, particularly in work domains where the amount of time which will elapse between the initiation of action and its realized outcome is measured in seconds and there will be little opportunity to act in any meaningful way to mitigate, correct, or redirect the course of events once they begin.

People are uncomfortable with uncertainty and tend to dismiss, delay, or assume, but engaging the unknown is a trainable skill, and the key is to act early (Mooney, 2020). For example, when road crews respond in hurricane conditions, to high water over a road, there are all types of unknowns or "UNs". You can probe many of the unknowns (cannot tell how deep), unclear (lack experience with water over road), uncertain (how fast water is flowing/how powerful the current), unseen (muddy, nighttime), uncontrollable (how much water is running under the road), and unstable (rapid changes – the road was fine driving to site but too deep when driving out). We teach methods to investigate and understand what is happening: look for swirls, look for water running in different directions, and share how people are commonly surprised; current sweeps you off your feet.

In addition, probing the unknown can take the form of questions: "How could we be surprised?" "What's making me uneasy?" "What can I see, and more importantly what can I NOT see?" "What do I know and NOT know?" "What's different?" "What else could this be?" "What do I need to pay close attention to?" "Who else can help?" The aim is to broaden and deepen the perspective that not all is knowable; however, there are questions and actions that can help move things from the unknown to known.

Finally, those who practice resilience engineering anticipate general surprise. We do not know exactly what will happen but we do know that in high tempo, high work load situations, something will happen. One strategy is to send in extra resources to "float"; for example, extra leadership without specific role/purpose to offer local help as needed during storm response. Another version of this is "dropping in an expert", a person with deep knowledge – maybe a retiree – to colocate with the overloaded team.

5 Monitor Weak Signals, Things Taking a Turn for the Worse

Sometimes surprise sneaks up on us.

Certain words and expressions ("I've never seen…" "We've never done…" "We're going to go slow…" "We don't have time to…" "worse than," "not sure," "maybe," "probably"), gestures or expressions (furrowed brow, nose and forehead scrunched up, widened eyes, gaping mouth, hand on chin or head), and feelings (tense, tightness in chest or stomach) reveal that we are sensing uncertainty or surprise. Training people to recognize these signs in ourselves and others can trigger us out of automaticity and into noticing a situation is changing which gives us the space to pause, reflect, and ask for help.

Per Salas et al., expertise-based intuition (basis for anticipation) is built through deliberate, guided practice including immediate pauses to monitor, reflect on, and fine-tune work (see AAR below) (Salas et al., 2010). Klein uses cognitive task analysis to mine cues noticed by experts then trains novices to use these cues to recognize increasing uncertainty, heightening risk, or signs trouble might be coming (Kahneman & Klein, 2009).

We can monitor for patterns of failure and for factors to combine, or "stack up." For example, we can train that if three or more situations emerge that increase difficulty in accomplishing work, pause, reassess, and get help. The pattern of "3", while not proven, has appeared in various industries (aviation 2 problems, troubleshoot…3 problems, land the plane) and the author's own experience investigating serious events. System components are interdependent and when stretched, approach boundaries of failure due to a reduction of human capacity to deal with challenges. This example is from a recent incident: he was a new worker, we were working on a very steep slope near a busy road, and it had been raining.

Other times, surprise is sudden and all we can do is respond.

6 Respond to Emergent Risk and Uncertainty

When uncertain or ambiguous situations are escalated, the role of leader goes beyond decision making. Leaders question, cross-check, and challenge. Leaders bring other experts into the conversation to probe risks, ensuring that all have a voice *before* leaders offer their own opinions. If the leader speaks too early, they risk closing down inputs of others. The safety team can select participants and invite people with specific expertise (e.g., the mechanic or equipment supplier), those who will bring a different perspective, then facilitate risk exploration:

- How could we fail?
- How could we be surprised?
- If something goes wrong, what's most likely to happen? How will it happen?
- If this is similar to…what's different that could make it harder?
- What must go right?
- What would help us do this safer?
- At the end, ask each person "what are your concerns?"

Finally, replan. For higher risk situations, insert hold points (when you get to this point, let us get back together and …) and off-ramps (if this happens, stop and …). The key is noticing that the situation has changed – using increasing uncertainty as a signal – then purposefully changing mode of operation: increasing vigilance, questioning, communication, and collaboration.

7 Learn from Surprise

The experience of surprise means our models of the world have broken down. This affords the opportunity to update our mental models to potentially avoid even more severe surprises in the future. Efforts to learn from surprise have included the initiation of standardized after-action reviews (AARs). This method, adapted from the military after-action review style has, though its early successes, achieved it principle goal of facilitating work-model updates both within the work team (between its members) and between each team and the larger organization as AARs can, and often or now shared across the company.

This version of an AAR is designed to be lightweight/low-overhead, thereby reducing the burden associated with its completion. As currently instituted, the AAR askes the practitioners to consider eight facets of the surprise experience, each with its own question:

1. What happened?
2. What was expected to happen?
3. What surprised us?
4. What prevented this event from being worse than it was? (event investigation)
5. Contributing factors? (event investigation)
6. What went well and why?
7. What can be improved and how?
8. What did we learn that would help others?

Earlier, we shared the characteristics of surprise for situations where a person was truck or almost struck by a limb or tree; we expand on our analysis of responses to "What surprised us?" and share how we can apply this to managing risk. Workers were surprised by how far the piece of wood flew, the direction it fell, and how fast the limb or tree fell. Saw operators were surprised that a person, often a new worker, walked into the fall zone after cutting began. The ground person was often surprised that the saw operator cut another branch when they thought the saw operator was finished (a number of incidents occurred when cutting the last branch). There were scale surprises such as a small twig taking down a power line or doing significant harm. There were hidden surprises such as decay in the tree when the tree looked healthy from the ground.

Identifying patterns of surprise supports workers better anticipating how they can be surprised and enables designing specific strategies such as practicing better team situation awareness (making actions predictable) especially at transition points in work and performing drills demonstrating the damage a small branch can do when dropped from height. Returning to Resilience Engineering fundamental tenant "surprise will happen" supports conversations about expecting the unexpected (this is comment is made frequently in leadership/worker conversations). In the case of struck-by risk, we teach to expect a new person to do something unexpected and expect, trees, especially Ash trees which are infected with Emerald Ash Borer, to fail in unexpected ways.

> …made the cut and the piece fell as intended, bouncing off of the dead Ash tree stem a few feet below the cut. As the limb made contact, the Ash tree failed, breaking approximately 11 feet from the base and fell across a single phase and neutral, bringing both wires to the ground and breaking the pole. What surprised us? The Ash tree failing as it did from such little additional strain due to rigging. Another stem, on the same tree, was removed a few days prior; the other stem was larger, 24 inches diameter at breast height, did not show any signs of rot or decay, and was removed without issue, in almost the exact same manner as the stem that failed.

Increased leadership empathy was an unexpected benefit of probing how people were surprised after events.

8 Summary

Early steps in shifting views on safety to embrace variability and accept surprise focus on noticing that safety is actively accomplished as part of everyday work. Efforts support noticing signs of emerging risk and increasing uncertainty (listen to your gut), then putting structure in place to manage work differently: bringing diversity to sensemaking and uncovering trade-off decisions. Both are largely accomplished through asking different questions, many examples of questions were provided in this chapter.

The next steps involve becoming more skilled at dealing with uncertainty and – even though this may sound like an oxymoron – become more consistent in managing critical variabilities. This can be accomplished through methods to deepen knowledge and build observation skills. Possible strategies to employ: bring experts (who likely do not know each other now) together to share experiences, tips, and techniques. This will build relationships to increase reciprocity. Video narration of trees we walk away from use these videos for training. Gather data for events and close calls to help us understand how things come together to create a step change in risk.

References

Hollnagel, E., Woods, D. D., & Leveson, N. (2006). *Resilience engineering: Concepts and precepts*. CRC Press.

ISHN. (2020, November 5). Top 25 most dangerous jobs in the United States. *Industrial Safety and Hygiene News*. https://www.ishn.com/articles/112748-top-25-most-dangerous-jobs-in-the-united-states.

Kahneman, D., & Klein, G. (2009). Conditions for intuitive expertise: A failure to disagree. *The American Psychologist, 64*(6), 515–526. https://doi.org/10.1037/a0016755.

Lynam, T., & Fletcher, C. (2015). Sensemaking: A complexity perspective. *Ecology and Society, 20*(1) http://www.jstor.org/stable/26269751.

Mooney, L. (2020, July). One reason we fail: The "Uns." Be Highly Reliable. https://www.behighlyreliable.com/read-more.

Salas, E., Rosen, M. A., & Diaz Granados, D. (2010). Expertise-based intuition and decision making in organizations. *Journal of Management, XX*(X), 1–31. https://doi.org/10.1177/0149206309350084.

Saxe, J. G. (1885). The blind men and the elephant. The poems of John Godfrey Saxe *(Highgate Edition)*, Houghton, Mifflin and Company.

Weber, D. E., MacGregor, S. C., Provan, D. J., & Rae, A. (2018). We can stop work, but then nothing gets done. *Safety Science., 108*, 149–160.

Weick, K., Sutcliffe, K., & Obstfeld, D. (2005). Organizing and the process of Sensemaking. *Organization Science, 16*(4), 409–421. http://www.jstor.org/stable/25145979.

Quo Vadis?

Erik Hollnagel and Christopher P. Nemeth

Contents

It is tempting, and perhaps even expected, to end a book as this by looking ahead, to speculating what should or could be done next to advance resilient performance. A premise for that is, of course, that something like resilient performance will still be needed. That this is so hardly needs any argument. The time when systems and societies could perform efficiently if everyone conscientiously followed simple rules and procedures ended at least a century ago, if not much earlier. Machines and systems are today non-trivial rather than trivial (von Foerster & Poerksen, 2002), in the sense that the transformations that link causes and effects are unknown rather than known. Since systems – and individuals – therefore must perform under conditions that are incompletely known, hence partly unpredictable, they must be able to adjust what they do in order to succeed. Any proposal for how something should or could be done – for Work-as-Imagined – therefore implies a set of assumptions about what the conditions will be about the World-as-Imagined.

On the positive side, people and systems still by and large perform well enough and are stable enough for societies to function, even during major disruptions. There is, therefore, ample evidence of what we may call resilient performance. The question is how we can understand what goes on, given that it by definition cannot be trivial, and how we can sustain and improve that. There has certainly been some progress since the heydays of human factors and behaviour-based safety, but much remains to be done. Performance can be resilient when there are recognisable patterns in how systems perform, but we need to understand both how these patterns

E. Hollnagel
Macquarie University, Sydney, Australia

C. P. Nemeth (✉)
Applied Research Associates, Inc., Albuquerque, NM, USA
e-mail: cnemeth@ara.com

C. P. Nemeth, E. Hollnagel (eds.), *Advancing Resilient Performance*,
https://doi.org/10.1007/978-3-030-74689-6_12

emerge and how people recognise and respond to them. We need to understand better how we come to accept certain assumptions about what Work-as-Imagined is and understand better how the regularity of Work-as-Done is established. We need to know what is required for a system to perform acceptably well now and in the future. To do so, it will be more important to ask the right questions than to hunt for improved answers to the questions that we uncritically – and often unsuccessfully – have tried to answer in the past.

Reference

Von Foerster, H., & Poerksen, B. (2002). *Understanding systems: Conversations on epistemology and ethics*. Carl-Auer-Systeme Verlag.

Printed in Great Britain
by Amazon